ASE Test Preparation Series

Truck Equipment

Installation and Repair
(Test E1)

DELMAR
CENGAGE Learning

Australia • Brazil • Japan • Korea • Mexico • Singapore • Spain • United Kingdom • United States

**ASE Test Preparation Series
Truck Equipment Installation and Repair (Test E1)**

Vice President, Technology and Trades Professional Business Unit: Gregory L. Clayton

Director, Professional Transportation Industry Training Solutions: Kristen L. Davis

Editorial Assistant: Danielle Filippone

Development: Dawn Jacobson

Director of Marketing: Beth A. Lutz

Senior Marketing Manager: Jennifer Barbic

Senior Production Director: Wendy Troeger

Production Manager: Sherondra Thedford

Content Project Management and Composition: PreMediaGlobal

© 2013 Delmar Cengage Learning

ALL RIGHTS RESERVED. No part of this work covered by the copyright herein may be reproduced, transmitted, stored or used in any form or by any means graphic, electronic, or mechanical, including but not limited to photocopying, recording, scanning, digitizing, taping, Web distribution, information networks, or information storage and retrieval systems, except as permitted under Section 107 or 108 of the 1976 United States Copyright Act, without the prior written permission of the publisher.

> For product information and technology assistance, contact us at
> **Cengage Learning Customer & Sales Support, 1-800-354-9706**
> For permission to use material from this text or product,
> submit all requests online at **www.cengage.com/permissions**
> Further permissions questions can be e-mailed to
> **permissionrequest@cengage.com**

Library of Congress Control Number: 2012930753

ISBN-13: 978-1-4354-3935-1

ISBN-10: 1-4354-3935-X

Delmar
5 Maxwell Drive
Clifton Park, NY 12065-2919
USA

Cengage Learning is a leading provider of customized learning solutions with office locations around the globe, including Singapore, the United Kingdom, Australia, Mexico, Brazil, and Japan. Locate your local office at: **international.cengage.com/region**.

Cengage Learning products are represented in Canada by Nelson Education, Ltd.

For more information on transportation titles available from Delmar, Cengage Learning, please visit our website at **www.trainingbay.cengage.com**.

For your lifelong learning solutions, visit **delmar.cengage.com**.

Visit our corporate website at **www.cengage.com**.

Notice to the Reader
Publisher does not warrant or guarantee any of the products described herein or perform any independent analysis in connection with any of the product information contained herein. Publisher does not assume, and expressly disclaims, any obligation to obtain and include information other than that provided to it by the manufacturer. The reader is expressly warned to consider and adopt all safety precautions that might be indicated by the activities described herein and to avoid all potential hazards. By following the instructions contained herein, the reader willingly assumes all risks in connection with such instructions. The publisher makes no representations or warranties of any kind, including but not limited to, the warranties of fitness for particular purpose or merchantability, nor are any such representations implied with respect to the material set forth herein, and the publisher takes no responsibility with respect to such material. The publisher shall not be liable for any special, consequential, or exemplary damages resulting, in whole or part, from the readers' use of, or reliance upon, this material.

Printed in the United States of America
1 2 3 4 5 6 7 16 15 14 13 12

Contents

TL 230 .T78 2013
Truck equipment

Preface iv

Section 1 The History and Purpose of ASE 1

Section 2 Take and Pass Every ASE Test 3
- How Are the Tests Administered? 3
- Who Writes the Questions? 3
- Objective Tests 4
- Preparing for the Exam 5
- During the Test 5
- Your Test Results! 6

Section 3 Types of Questions on an ASE Exam 7
- Multiple-Choice Questions 7
- EXCEPT Questions 7
- Technician A, Technician B Questions 8
- Most Likely Questions 9
- LEAST LIKELY Questions 9
- Summary 9
- Testing Time Length 10

Section 4 Overview of the Task List 11
- Truck Equipment Installation and Repair (Test E1) 11
- Task List and Overview 11
    - A. Frame Preparation (11 Questions) 11
    - B. Suspension Systems (5 Questions) 15
    - C. Driveshaft Preparation (5 Questions) 18
    - D. Body and Equipment (24 Questions) 20

Section 5 Sample Test for Practice 25
- Sample Test 25

Section 6 Additional Test Questions for Practice 33
- Additional Test Questions 33

Section 7 Appendices 59
- Answers to the Test Questions for the Sample Test Section 5 59
- Explanations to the Answers for the Sample Test Section 5 60
- Answers to the Test Questions for the Sample Test Section 6 68
- Explanations to the Answers for the Sample Test Section 6 70

Glossary 91

Preface

Delmar, a part of Cengage Learning, is very pleased that you have chosen our ASE Test Preparation Series to prepare yourself for the Truck Equipment Installation and Repair (Test E1) ASE examination. This guide is designed to introduce you to the task list for the Truck Equipment Installation and Repair (Test E1) test you are preparing to take, give you an understanding of what you are expected to be able to do in each task, and take you through sample test questions formatted in the same way the ASE tests are structured.

If you have a basic working knowledge of the discipline you are testing for, you will find the Delmar Learning's ASE Test Preparation Series to be an excellent way to understand the "must know" items to pass the test. This book is not a textbook. Its objective is to prepare the technician who has the requisite experience and schooling to challenge ASE testing. It cannot replace the hands-on experience or the theoretical knowledge required by ASE to master vehicle repair technology. If you are unable to understand more than a few of the questions and their explanations in this book, it could be that you require either more shop-floor experience or further study.

This book begins with an item-by-item overview of the ASE Task List, with explanations of the minimum knowledge you must possess to answer questions related to the task. Following that, there are two sets of sample questions, followed by an answer key to each test and an explanation of the answers to each question. A few of the questions are not strictly ASE format but were included because they help teach a critical concept that will appear on the test. We suggest that you read the complete task list overview before taking the first sample test. After taking the first test, score yourself and read the explanation to any questions that you were not sure about, including the questions you answered correctly. Each test question has a reference back to the related task or tasks that it covers. This will help you to go back and read over any area of the task list that you are having trouble with. Once you are satisfied that you have all of your questions answered from the first sample test, take the additional test and check it. If you pass these tests, you will do well on the ASE test.

Our Commitment to Excellence

Delmar Cengage Learning has sought out the best technicians in the country to help with the development of this first edition of the Truck Equipment Installation and Repair (Test E1) ASE Test Preparation Guide.

About the Author

Peter Taskovic has been in the automotive and truck repair industry and 2nd-stage manufacturing business for over 30 years. He has built and run everything from front-engine dragsters to 40-metric ton truck-mounted cranes. Peter has been ASE certified for over 25 years and currently holds Master Level Certificates in Automotive, Truck Repair, Body Refinishing, and Truck Equipment Repair as well as individual Alternative Fuels and Service Consultant certificates. He is also certified as a CSA Fuel System Inspector for CNG Vehicles. Peter played in integral part of the team that helped write the E Series Tests for ASE. Currently he holds the position of Manager of Technical Services at Auto Truck Group, LLC, where he has been employed for 25 years.

Thanks for choosing Delmar Learning ASE Test Preparation Series. All of the writers, editors, and Delmar staff have worked very hard to make this test preparation guide second to none. We know you are going to find this book accurate and easy to work with. It is our objective to constantly improve our products at Delmar by responding to feedback.

If you have any questions concerning the books in this series, please visit us on the Web at http://www.trainingbay.com

1 The History and Purpose of ASE

ASE began as the National Institute for Automotive Service Excellence (NIASE). It was founded as a non-profit, independent entity in 1972 by a group of industry leaders with the single goal of providing a means for consumers to distinguish between incompetent and competent technicians. It accomplishes this goal by testing and certifying repair and service professionals. From this beginning, it has evolved to be known simply as ASE (Automotive Service Excellence) and today offers more than 40 certification exams in automotive, medium/heavy-duty truck, collision, engine machinist, school bus, transit bus, parts specialist, automobile service consultant, and other industry-related areas. There are now more than 400,000 professionals with current ASE certifications. These professionals are employed by new car and truck dealerships, independent garages, fleets, service stations, and franchised service facilities, to name a few. ASE continues its mission by providing information that helps consumers identify repair facilities that employ certified professionals through its Blue Seal of Excellence Recognition Program. Shops having a minimum of 75 percent of their repair technicians which are ASE certified, and meet other criteria, can apply for and receive the Blue Seal of Excellence Recognition from ASE.

ASE recognized that educational programs serving the service and repair industry also needed a way to be recognized as having the faculty, facilities, and equipment required to provide quality education to students wanting to become service professionals. Through the combined efforts of the ASE, the industry, and leaders in education, the non-profit National Automotive Technicians Education Foundation (NATEF) was founded to evaluate and recognize training programs. Today, more than 2,000 programs are ASE certified under standards set by the service industry. ASE/NATEF also has a certification-of-industry-(factory) training program known as Continuing Automotive Service Education (CASE). CASE recognizes training programs offered by replacement parts manufacturers as well as vehicle manufacturers.

ASE's certification testing is administered by American College Testing (ACT). Strict standards of security and supervision at the test centers ensure that the technician who holds the certification has earned it. Additionally, ASE certification requires the candidate to be able to demonstrate two years of work experience in the field before certification. Test questions are developed by industry experts who are actually working in the field. Details on how the test is developed and administered are provided in Section 2. The certification is valid for five years and can be recertified by retesting. ASE issues a jacket patch, a certificate, and a wallet card to certified technicians and makes signs available to facilities that employ ASE-certified technicians. This is to enable consumers to recognize certified technicians.

You can contact ASE at:

National Institute for Automotive Service Excellence
101 Blue Seal Drive S.E.
Suite 101
Leesburg, VA 20175
Telephone: 703-669-6600
Fax: 703-669-6123
http://www.ase.com

2 Take and Pass Every ASE Test

Participating in an Automotive Service Excellence (ASE) voluntary certification program gives you a chance to show your customers that you have the know-how needed to work on today's modern vehicles. The ASE certification tests allow you to judge your skills and knowledge against the automotive service industry's standards for each specialty area.

If you are the "average" automotive technician taking this test, you are in your mid-30s and have not attended school for about 15 years. That means you probably have not taken a test in many years. Some of you, on the other hand, may have attended college or taken post-secondary education courses and may be more familiar with taking tests and with test-taking strategies. There is, however, a difference between the educational tests you may be accustomed to and the ASE test you are preparing to take.

How Are the Tests Administered?

ASE administers its certification exams utilizing a Computer Based Testing (CBT) methodology. The CBT exams are administered at test centers across the nation.

While it is always recommended that you refer to the ASE website (*www.ase.com*) for the latest data regarding testing registration and exam dates, below is an overview of the available testing windows. CBT exams will be available four times annually, for two-month windows, with a month of no-testing in between each testing window.

- January/February—Winter CBT Testing Window
- April/May—Spring CBT Testing Window
- July/August—Summer CBT Testing Window
- October/November—Fall CBT Testing Window

Who Writes the Questions?

ASE test questions are written by service industry experts in the area being tested. Each area will have its own technical experts. Questions are entirely job related. They are designed to test the skills you need to be a successful technician. Theoretical knowledge is important and necessary to answer the questions, but ability to apply that knowledge is the basis of ASE test questions.

Each question has its roots in an ASE "item-writing" workshop, where service representatives from automobile manufacturers (domestic and import), aftermarket parts and equipment manufacturers, working technicians, and vocational educators meet to share ideas and translate them into test questions. Each test question written by these experts must survive review by all members of the group.

The questions are written to deal with the practical application of soft skills and knowledge of systems experienced by technicians in their day-to-day work. All questions are pretested and quality-checked by a national sample of technicians.

Those questions that meet ASE standards of quality and accuracy are included in the scored sections of the tests; the "rejects" are sent back to the drawing board or discarded altogether.

Each certification test is made up of between 40 and 80 multiple-choice questions.

Note, however, that each test could contain additional questions that are included for statistical research purposes only. Your answers to these questions will not affect your score, but since you do not know which ones they are, you should answer all questions in the test. The once-in-five-year recertification test will cover the same content areas as those listed in the preceding text. However, the number of questions in each content area of the recertification test will be reduced by about one-half.

Using multiple criteria, including cross-sections by age, race, and other background information, ASE is able to guarantee that a question does not bias for or against any particular group. A question that shows bias toward any particular group is discarded.

Objective Tests

A test is called an objective test if the same standards and conditions apply to everyone taking the test and there is only one correct answer to each question.

Objective tests primarily measure your ability to recall information. A well-designed objective test can also test your ability to understand, analyze, interpret, and apply your knowledge. Objective tests include true–false, multiple-choice, fill-in-the-blank, and matching questions. ASE's tests consist exclusively of four-part multiple-choice objective questions.

The following are some strategies that may be applied to taking your tests.

Before beginning an objective test, quickly look over the test to determine the number of questions, but do not try to read through all the questions. In an ASE test, there are usually between 40 and 80 questions, depending on the subject. Read through each question before marking your answer. Answer the questions in the order they appear on the test. Leave those questions blank that you are not sure of and move on to the next question. You can return to those unanswered questions after you have finished the others. They may be easier to answer at a later time after your mind has had additional time to consider them at a subconscious level. In addition, you might find information in other questions that will help you recall the answers to some of the unanswered ones.

Do not be obsessed by the apparent pattern of responses. For example, do not be influenced by a pattern like **D, C, B, A, D, C, B, A** on an ASE test.

There is also a lot of folk wisdom about taking objective tests. For example, there are those who would advise you to avoid response options that use words such as *all, none, always, never, must,* and *only,* to name a few. This, they claim, is because nothing in life is exclusive. They would advise you to choose response options that use words that allow for some exception, such as *sometimes, frequently, rarely, often, usually, seldom,* and *normally.* They would also advise you to avoid the first and last option (**A** or **D**) because test writers, they feel, are more comfortable if they put the correct answer in the middle (**B** or **C**) of the list of choices. Another recommendation often offered is to select the option that is either shorter or longer than the other three choices because it is more likely to be correct. Some would advise you to never change an answer since your first intuition is usually correct.

Although there may be a grain of truth in this folk wisdom, ASE test writers try to avoid them, and so should you. There are just as many **A** answers as there are **B** answers, and just as many **C** answers as **D** answers. As a matter of fact, ASE tries to balance the answers at about 25 percent per choice **A, B, C,** and **D.** There is no intention to use "tricky" words, such as those outlined previously. Put no credence in the opposing words "sometimes" and "never," for example.

Multiple-choice tests are sometimes challenging because there are often several choices that may seem possible, and it may be difficult to decide on the correct choice.

The best strategy, in this case, is to first determine the correct answer before looking at the options. After arriving at the answer you have worked out, you should still examine the options given to make sure that none seems more correct than yours. If you do not know or are not sure of the answer, read each option carefully and try to eliminate those options that you know are incorrect. That way, you can often arrive at the correct choice through a process of elimination.

If you have gone through all the test questions and you still do not know the answer to some of the questions, then guess. Yes, guess. You then have at least a 25 percent chance of being correct. If you leave the question blank, you have no chance. Your score is based on the number of questions answered correctly.

Preparing for the Exam

The main reason we have included so many sample and practice questions in this guide is, simply, to help you learn what you know and what you don't know. We recommend that you work your way through each question in this book. Before doing this, carefully look through Section 3: It contains a description and explanation of the types of questions you'll find in an ASE exam.

Once you understand what the questions will look like, move to the sample test questions (Section 5). Read the explanations (Section 7) to the answer for each question, and if you don't feel you understand the reasoning for the correct answer, go back and read the overview (Section 4) for the task that is related to that question. If you still don't feel you have a solid understanding of the material, identify a good source of information on the topic, such as a textbook, and do some more studying.

After you have completed all the sample test items and reviewed your answers, move to the additional questions (Section 6). This time, answer the questions as if you were taking an actual test. Do not use any reference or allow any interruptions so as to get a feel for how you will do in an actual test. Once you have answered all the questions, grade your results using the answer keys in Section 7. For every question to which you gave an incorrect answer, study the explanations to the answers and/or the overview of the related task areas (Section 4). Try to determine the root cause for your missing the question. The easiest thing to correct is learning the correct technical content. The hardest thing to correct is the behavior that leads you to the incorrect answer. If you knew the information but still got it incorrect, there is a behavior problem that will need to be corrected. An example would be reading too quickly and skipping over words, which affect your reasoning. If you can identify what you did that caused you to answer the question incorrectly, you can eliminate that cause and improve your score. Here are some basic guidelines to follow:

- Focus your studies on those areas you are weak in.
- Be honest with yourself while determining if you understand something.
- Study often but in short periods of time.
- Remove yourself from all distractions while studying.
- Keep in mind that the goal of studying is not just to pass the exam: The real goal is to learn!
- Prepare physically by getting a good night's rest before the test and have meals that provide energy but do not cause discomfort.
- Arrive early at the test site to avoid long waits as test candidates check in and to allow all the time available for your tests.

During the Test

When taking a CBT exam, as soon as you are seated in the testing center, you will be given a brief tutorial to acquaint you with the computer-delivered test, prior to taking your certification exam(s). The CBT exams allow you to only select one answer per question. You can also change your answers as many times as you like. When you select a second answer choice, the CBT will automatically unselect your first answer choice. If you want to skip a question to return to later, you can utilize their "flag" feature, which will allow you to quickly identify and review questions whenever you are ready. Prior to completing your exam, you will also be provided with an opportunity to review your answers and address any unanswered questions.

If you have finished answering all the questions on a test and have time remaining, go back and review the answers of those questions that you were not sure of. You can often catch careless errors by using the remaining time to review your answers. Carefully check your answer sheet for blank answer blocks or missing information.

At practically every test, some technicians will invariably finish ahead of time and complete their tests long before the final call. Some technicians may be taking a recertification test and others may be taking fewer tests than you. Do not let them distract or intimidate you.

It is not wise to use less than the total time that you are allotted for a test. If there are any doubts, take the time for review. Any product can usually be made better with some additional effort. A test is no exception.

Your Test Results!

You can gain a better perspective about tests if you know and understand how they are scored. ASE's tests are scored by a non-partial, unbiased organization having no vested interest in ASE or in the automotive industry.

Each question carries the same weight as any other question. For example, if there are 50 questions, each is worth 2 percent of the total score. The passing grade is 70 percent. That means you must correctly answer 35 of the 50 questions to pass the test.

The test results can tell you (1) where your knowledge equals or exceeds that needed for competent performance or (2) where you might need more preparation.

Your ASE test score report is divided into content areas and will show the number of questions in each content area and how many of your answers were correct. These numbers provide information about your performance in each area of the test. However, because there may be a different number of questions in each content area of the test, a high percentage of correct answers in an area with few questions may not offset a low percentage in an area with many questions.

It should be noted that one does not "fail" an ASE test. The technician who does not pass is simply told "more preparation needed." Though large differences in percentages may indicate problem areas it is important to consider how many questions were asked in each area. Since each test evaluates all phases of the work involved in a service specialty, you should be prepared in each area. A low score in one area could keep you from passing an entire test.

There is no such thing as average. You cannot determine your overall test score by adding the percentages given for each task area and dividing by the number of areas. It doesn't work that way because generally the number of questions in each task area is not the same. A task area with 20 questions, for example, counts more toward your total score than a task area with 10 questions.

Your test report should give you a good picture of your results and a better understanding of your strength and weaknesses for each task area.

If you fail to pass the test, you may take it again at any time it is scheduled to be administered. You are the only one who will receive your test score. Test scores will not be given over the telephone by ASE nor will they be released to anyone without your written permission.

3 Types of Questions on an ASE Exam

ASE certification tests are often thought of as being tricky. They may seem to be tricky if you do not completely understand what is being asked. The following examples will help you recognize certain types of ASE questions and avoid common errors.

Most initial certification tests are made up of 40–80 multiple-choice questions. Multiple-choice questions are an efficient way to test knowledge. To answer them correctly, you must think about each choice as a possibility, and then choose the one that best answers the question. To do this, read each word of the question carefully. Do not assume you know what the question is about until you have finished reading it.

About 10 percent of the questions on an actual ASE exam will use an illustration. These drawings contain the information needed to correctly answer the question. The illustration must be studied carefully before attempting to answer the question. Often, technicians look at the possible answers and then try matching the answers to the drawing. Always, however, do the opposite: match the drawing to the answers. When the illustration is showing an electrical schematic or another system in detail, look over the system and try to figure out how the system works before you look at the question and the possible answers.

Multiple-Choice Questions

The most common type of question used on ASE tests is the multiple-choice question. This type of question contains three "distracters" (incorrect answers) and one "key" (correct answer). When the questions are written, effort is made to make the distracters plausible to draw an inexperienced technician to one of them. This type of question gives a clear indication of the technician's knowledge. If you encounter a question that you are unsure of, reverse engineer it by eliminating the items that it cannot be. Consider the following example:

1. Which of the following would be the result of an electrical short to ground before the load?
    A. Reduced current flow
    B. Circuit protection device opens
    C. Circuit operates normally
    D. Voltage drop across the load is increased   (A7)

Answer A is incorrect. The current flow would be greatly increased, which would cause the circuit protection device to open.
Answer B is correct. A short to ground before the load would cause the electrical flow to run straight to ground and bypass the load in the circuit. This would cause the circuit resistance to drop very low, which would cause a dramatic increase in current flow, which would open the circuit protection device.
Answer C is incorrect. The circuit would not operate normally due to the shorted path to ground before the load.
Answer D is incorrect. The load in the circuit would not have any voltage applied to it due to the shorted path to ground before the load.

EXCEPT Questions

Another type of question used on ASE tests has answers that are all correct except one. The correct answer for this type of question is the answer that is incorrect. The word "EXCEPT" will always be in capital letters. You must identify which of the choices is the incorrect answer. If you read too quickly through the question, you may overlook what the question is asking and answer the question with the first correct statement. This will make your answer incorrect. An example of this type of question and the analysis is as follows:

1. All of the following electrical tools can be used as electrical conductors in a circuit EXCEPT:
   A. Copper
   B. Aluminum
   C. Rubber
   D. Gold   (A2)

Answer A is incorrect. Copper is an excellent conductor that is widely used to connect electrical circuits.
Answer B is incorrect. Aluminum is an excellent conductor that is often used to make electrical terminals.
Answer C is correct. Rubber is classified as an insulator, which resists electrical flow.
Answer D is incorrect. Gold is an excellent conductor and is sometimes used on the tips of critical electrical terminals to improve the quality of the connection.

Technician A, Technician B Questions

The type of question that is most popularly associated with an ASE test is the "Technician A says . . . Technician B says . . . Who is correct?" type. In this type of question, you must identify the correct statement or statements. To answer this type of question correctly, you must carefully read each technician's statement and judge it on its own merit to determine if the statement is true.

Sometimes, this type of question begins with a statement about some analysis or repair procedure. This is often referred to as the stem of the question and provides the setup or background information required to understand the conditions on which the question is based. This is followed by two statements about the cause of the concern, proper inspection, identification, or repair choices. You are asked whether the first statement, the second statement, both statements, or neither statement is correct. Analyzing this type of question is a little easier than the other types because there are only two ideas to consider, although there are still four choices for an answer.

Technician A, Technician B questions are really double true-or-false questions. The best way to analyze this type of question is to consider each technician's statement separately. Ask yourself, is A true or false? Is B true or false? Then select your answer from the four choices. An important point to remember is that an ASE Technician A, Technician B question will never have Technician A and B directly disagreeing with each other. That is why you must evaluate each statement independently.

An example of this type of question follows:

1. The headlights on a truck are very dim on high and low beams. Technician A says that a loose common ground connection for the headlights could cause this problem. Technician B says that a blown headlight fuse could cause this problem. Who is correct?
   A. A only
   B. B only
   C. Both A and B
   D. Neither A nor B   (E1)

Answer A is correct. A loose common ground connection for the headlights could cause this problem. The technician would notice that the voltage drop on the headlights would be less than system voltage. The problem could be isolated by checking the voltage drop in the power and ground side of the circuit.
Answer B is incorrect. A blown headlight fuse would cause the headlights to be totally inoperative.
Answer C is incorrect. Only Technician A is correct.
Answer D is incorrect. Technician A is correct.

Most Likely Questions

Most likely questions are somewhat difficult because only one choice is correct while the other three choices are nearly correct. An example of a most likely question is as follows:

1. An operator complains that the backup alarm sounds low and weak. Which of these is the most likely cause?
   A. An open backup switch
   B. A stuck-closed backup switch
   C. A blown backup light bulb
   D. A burnt contact on the backup alarm relay (B12)

Answer A is incorrect. An open backup switch would cause the alarm to be completely inoperative.
Answer B is incorrect. A stuck-closed backup switch would cause the alarm to sound continuously.
Answer C is incorrect. A blown backup light bulb would not likely cause any issues with the backup alarm.
Answer D is correct. The relay for the backup alarm delivers the required current needed to operate the alarm. If the "load side" contacts get burned or corroded, then current flow is decreased due to the increased electrical resistance in the circuit.

LEAST LIKELY Questions

Notice that in the most likely questions, there is no capitalization. This is not so with LEAST LIKELY-type questions. For this type of question, look for the choice that would be the LEAST LIKELY cause of the described situation. Read the entire question carefully before choosing your answer.
An example is as follows:

1. The fuse for the heater and AC blows every time the blower motor is turned on. Which of the following is the LEAST LIKELY cause?
   A. A loose ground at the blower motor
   B. A short to ground in the blower power circuit
   C. A shorted blower motor
   D. A locked up blower motor (F4)

Answer A is correct. A loose ground at the blower motor would cause the blower motor to run slower or not at all. It would not cause the fuse to blow because the circuit electrical resistance would be higher, which would cause current flow to decrease.
Answer B is incorrect. A short to ground before the load will cause a circuit protection device to open.
Answer C is incorrect. A shorted blower motor could cause the fuse to blow each time the blower is turned on.
Answer D is incorrect. A locked up blower motor would cause the electrical current to spike and result in the fuse blowing.

Summary

There are no four-part multiple-choice ASE questions having "none of the above" or "all of the above" choices. ASE does not use other types of questions, such as fill-in-the-blank, completion, true-false, word-matching, or essay. ASE does not require you to draw diagrams or sketches. If a formula or chart is required to answer a question, it is provided for you. There are no ASE questions that require you to use a pocket calculator.

Testing Time Length

Each individual ASE CBT exam has a fixed time limit. Individual exam times will vary based upon exam area, and will range anywhere from a half hour to two hours. You will also be given an additional 30 minutes beyond what is allotted to complete your exams to ensure you have adequate time to perform all necessary check-in procedures, complete a brief CBT tutorial, and potentially complete a post-test survey. You should be on time to ensure that you have all of the allocated time available. If you arrive late for a CBT test appointment, you will only have the amount of time remaining in your appointment.

Visitors are not permitted at any time. If you wish to leave the testing room for any reason, you must first ask permission. Even if you finish your test early and wish to leave, you would be permitted to do so only during specified dismissal periods.

You should monitor your progress and set an arbitrary limit to how much time you will need for each question. This should be based on the number of questions you are attempting. It is suggested that you wear a watch because some facilities may not have a clock visible to all areas of the room.

4 Overview of the Task List

Truck Equipment Installation and Repair (Test E1)

The following section includes the task areas and task lists for this test and a written overview of the topics covered in the test.

The task list describes the actual work you should be able to do as a technician that you will be tested on by ASE. This is your key to the test and you should review this section carefully. We have based our sample test and additional questions on these tasks and the overview section will also support your understanding of the task list. ASE advises that the questions on the test may not equal the number of tasks listed; the task lists tell you what ASE expects you to know how to do and be ready to be tested on.

At the end of each question in the Sample Test and Additional Test Questions sections, a letter and number will be used as a reference to this section for additional study. Note the following example:

1. Prior to beginning the installation of a body on a chassis-cab, a technician should:
   A. Check all information provided in the chassis manufacturers Body Builder's Guides and Manuals
   B. Using only a tape measure should ensure accurate dimensions
   C. Check only the wheelbase dimension
   D. Check the dimensions in the Incomplete Vehicle Document (A1)

Answer A is correct. All accurate technical and dimensional information is in the Body Builder's Manual for the individual cab-chassis that a technician is working on.
Answer B is incorrect. Using only a tape measure in not accurate for this task.
Answer C is incorrect. While important, the wheelbase is only one of the important dimensions you will need to know.
Answer D is incorrect. The Incomplete Vehicle Document does not contain accurate dimensions.

Task List and Overview

A. Frame Preparation (11 Questions)

Task A1 **Verify wheelbase, cab-to-axle/cab-to trunnion measurements, and tandem axle spread.**

It is critical that all dimensions are known and validated prior to the installation of the body. This information is available in each chassis manufacturers' Body Builder's Manual. In the case of a single axle chassis, the dimensions will include wheelbase (WB), cab-to-axle (CA), cab-to-body (CB), cab-to-end of frame, and on occasions, bumper to back of cab, also known as BBC. On a tandem axle truck, wheelbase will be measured from the center of the rear duals to the center of the front axle. In light of the fact that the placement of body and equipment on the chassis will effect vehicle weight distribution and braking, these dimensions must be accurate and correct.

Task A2 Verify gross vehicle weight rating (GVWR) and gross axle weight ratings (GAWR).

Gross vehicle weight ratings and gross axle weight ratings are used to determine the maximum usage and application of the vehicle from a load bearing/carrying standpoint. These values will determine the safe operating range of the completed vehicle. These values can be found on a data plate attached to the truck cab, in the Body Builder's Manual, and in the Incomplete Vehicle Document, located in the truck cab at the factory after assembly of the truck chassis. Because these values effect steering performance, suspension performance and braking performance as outlined in the Federal Register under Title 49, Transportation, and the Motor Vehicle Safety Standards as published in Title 49.

Task A3 Verify frame width, height, and length.

While the generally accepted standard frame width in the United States and Canada is 34 inches, this is not always the case on the truck a technician might be working on. Many imported chassis do not comply with the 34 inch generally accepted principle. It is important, therefore, that a technician check this dimension both on the truck and in the Body Builder's Manuals. Frame length is also found in the Body Builder's Manual, and it can also be measured on the truck. Frame height is more difficult because the technician must allow for spring deflection. Original equipment truck chassis manufacturers use various recommended practices for calculating chassis height under specific loads. This is also found in the Body Builder's Manuals or other OEM Publications.

Task A4 Determine body/equipment layout locations.

A visual inspection of the chassis should take place prior to beginning any work. Discovering the location of chassis related components prior to the beginning of work will speed up the mounting process and provide a quality installation. Drawings should be provided to a technician from the engineering department, they should take time to familiarize themselves with the completed design. A technician must be aware of which components may be unavailable for relocation due to emissions requirements, operational requirements or other issues.

Task A5 Remove and relocate frame-mounted vehicle components/systems as necessary.

This task requires a good knowledge of fabrication techniques, welding, mounting hardware, frame piercing, as well as the electrical and hydraulic systems. Incorrectly moved or relocated chassis equipment can result in catastrophic events such as fires, explosions, and components failures.

The technician must have a good level of understanding of the legal requirements of the task about to be performed. On all trucks equipped with an EPA Certified Diesel Engines equipped with a Diesel Particulate Filter, many components of the exhaust system may not be relocated. This may also apply to air intake systems certified for EPA noise regulations, gas engine fuel systems that are equipped with evaporative canister units, and even anti-lock brake components that may be chassis mounted.

The relocation of other chassis related components like battery, battery boxes, high load circuit breaker boxes, chassis mounted fuel tanks; heated fuel water separators, heated air dryers, and other assorted components may generally be moved, within the OEM guidelines.

The first step in any chassis modification is to disconnect the vehicle batteries. This protects the on-board computer systems, the charging system, and is a recommended practice by almost every organization.

As mentioned earlier, knowledge in frame piercing is critical to this part of the job. A technician will use various methods to fabricate holes in the frame rail. This task may also require the removal of rivets and other types of fasteners. A technician will be expected to install holes in suitable locations in a frame rail. Holes should not be drilled in the top or bottom frame flange. Hole locations on the frame web generally cannot be within 2 inches from the frame radius but each chassis manufacturer may have their own requirements that will be published in the Body Builder's Manual.

Task A6   Repair, lengthen, shorten, or reinforce frame and frame members.

The frame is the backbone of the truck and must be treated with great respect. The frame is carrying the load, the body and equipment with great stress loads. It is critical that the technician, be knowledgeable about frame construction, material strength, welding techniques, and frame ratings.

Frame specifications include the following terms—Section Modulus, Yield Strength, Tensile Strength, and Resistance Bending Moment. Any time that a technician is working with a truck frame, they must be familiar with these terms.

Section Modulus (SM) is a representation of the volume of steel in a particular location in the frame. Section Modulus will not be consistent from the front of the frame to the back. It will change along the way. For example, the SM at the rear of a tapered frame truck will be smaller in dimension and volume than that section of the frame directly behind the cab. In most cases, published section modulus is the dimension located at the rear of the cab area of the frame. Likewise, frame section modulus at the front of the frame may also be smaller than other areas of the frame. This would be required knowledge when mounting a snow plow.

Section Modulus is a result of a calculation that includes frame thickness, web dimension, flange dimension and radius calculations at the point where the web of the frame "turns" into the flange of the frame. This calculation cannot be created by anyone other than the original equipment manufacturer. Some manufacturers provide a section modulus chart and graph that depicts the section modulus at various points along the frame, usually using the front axle of the truck as the data point. This dimension is also published by many body and equipment manufacturers as a "minimum" dimension for a particular body or piece of equipment. For example, a crane manufacturer will always publish a minimum section modulus requirement. Even certain lift gate manufacturers will publish minimum section modulus data requirement data.

Yield strength is a measurement of the strength of the steel used by the frame manufacturer. It is not to be confused with tensile strength. While each represents a similar form of rating the physical properties of steel, they are slightly different. Yield strength is defined as the point at which a material will take on a permanent change in shape. For example, when applying a force to a section of steel, the steel will flex. The elasticity of the steel will allow it to return to the original strength. However, at some point, the steel will bend and take on a new shape. The point at which the steel permanently deforms is the yield strength. This is generally measured in pounds per square inch (psi).

Tensile strength is a similar measurement. It is the ultimate strength of material. This represents the maximum strength of a material when subjected to tension, compression, or shear. Most manufacturers use yield strength as the value of steel when providing specifications. It is rare to see the term value tensile strength used. Truck frames can be manufactured with various strength steel. There may be ratings from 34,000 psi yield to as high as 125,000 psi yield. In the case of 125,000 psi steel, it would be describing a heat treated frame typically found on a class eight tractor. Lower ratings, such as 34,000 psi to 55,000 psi would be found in lighter-duty straight trucks, typically class three thorough seven.

Truck frames can be produced using steel or aluminum. In each case, the tensile and yield strength values are measured in the same way.

The overall measure of frame strength is what is known as Resisting Bending Moment, or simply RBM. RBM is calculated by multiplying the yield strength by the section modulus. The result is measured in inch-pounds. The higher this value, the stronger the frame. A frame of higher yield strength and small section modulus can result in a light weight component that does not need reinforcement.

All of this is important information when modifying a truck frame in any way. Lengthening a truck frame, for example, could mean adding after-frame for a particular piece of equipment, or just to support the rear of the body. It could also mean lengthening the wheelbase. Shortening a frame could also mean shortening the after-frame or reducing the wheelbase of a truck chassis. When shortening or lengthening a wheelbase, many considerations must be made. Lengthening by relocating the rear suspension, and modifying the driveline, is one way of doing this. Cutting and adding a section of frame is another. In this case, the size, quality, and yield and tensile strength ratings must be taken into consideration. The RBM rating of the frame will be affected by the yield strength of the material selected to be used as the insert. The material of the lowest value will be used for the RBM calculation.

In cases where the frame has been modified, it may be required to add a reinforcement section. This also maybe required on standard frames that lack the section modulus for items such as a crane. In most cases, chassis manufacturers Body Builder's Manuals will illustrate the proper method of constructing and installing frame reinforcement. There are several ways to reinforce a frame. A full wrap C channel reinforcement is essentially a web, two flanges and radius that is manufactured to fit completely around the original frame. Fishplate reinforcement is a single plate that is attached to the web of the frame only. This can be on the inside of the frame, the outside of the frame, or both.

The inverted L section reinforcement is an L shaped piece of steel that covers the top flange and the web of the original frame. The lower L section reinforcement is just the opposite of the inverted L. It wraps around the lower frame flange and the web. The lower L can be more prone to corrosion as it can trap moisture between the frame web and the reinforcing section rail.

How these reinforcement plates are attached is also important. Use of grade eight nuts, bolts and washers is acceptable, and in some cases, reinforcements can be plug welded as well.

Task A7 Layout and drill mounting holes.

The nature of this task requires absolute precision when measuring for hole placement. If the placement of holes is due to the relocation of a rear axle assembly due to a change in wheelbase and the subsequent movement of the vehicle rear spring hanger locations, the placement of such holes will alter the differential pinion angle. This will cause irregular driveshaft/universal joint operating angles and could cause severe problems including driveline catastrophic failures and progressive damage to the vehicle transmission, underbody, and rear differential.

Task A8 Disconnect battery and chassis components that are sensitive to welding procedures.

Since the late 1980's, all commercial vehicles, both on-road and off-road, have been equipped with some type of computer system. This could be an engine control module, a transmission control unit, or an anti-lock brake system control module. It is absolutely necessary to protect these control modules when performing any welding tasks on a truck, body, or any attached unit, regardless of how small or fast the welding job is. The battery ground cable must be completely disconnected from the chassis and isolated to prevent accidental contact with a surface being welded.

Task A9 Inspect frame and frame members for cracks, breaks, distortion, elongated holes, looseness, and damage; determine needed repairs.

A careful inspection of the chassis/frame should be conducted prior to body mounting. If the truck is a used vehicle, it is important to determine if previous alterations have been performed correctly. If not, correction should be included in the body mounting process. A tape measure should be used to determine if the chassis is square and true. Inspection of cross members should include looking for previous welding, grinding for clearance, bending due to previous accidents, cracking due to stress, and hole drilling for equipment mounting. It is also of great importance to inspect for rust and other corrosion, especially if the chassis was located or stored in a "rust belt" environment. Rust penetration must be repaired prior to any body mounting procedure.

This is also a good time to inspect clearance of components. Fuel lines, brake lines/hoses, and wiring harnesses should be mounted with adequate clearance so as not to create potential damage from contact or heat.

Task A10 Inspect, install or repair frame, hangers, brackets, and cross members.

Problems discovered during inspection of frame, hangers, brackets or cross members must be repaired. When repairing a frame, it is important to understand the basics of welding and know the type of material being repaired. Cold rolled steel will require a different welding process than heat treated steel. Damaged brackets, such as fuel tank "J" brackets, must be carefully repaired should damage be discovered. In no case should any frame-mounted components and mounting brackets be welded directly to the frame.

B. Suspension Systems (5 Questions)

Task B1 — Preparation and Installation (3 Questions)

Task B1.1   Relocate suspension components as necessary.

Suspension components that require relocation will generally be a result of a wheelbase modification requirement. It some instances, a technician will relocate the suspension cross members, hangers, shackles, shock absorber mounts, and other devices by removing them entirely from the chassis, laying out a new location based on wheelbase dimension, pierce new mounting holes using a template, and remounting the components using new mounting hardware. In other cases, a technician will leave all components attached to the frame, section the frame, extend or shorten the frame within the wheelbase area, and weld the frame, with or without an extension, back together again. This will require reinforcing the spliced areas of the frame, carefully. In the case of a tapered after frame, it may not be possible to relocate the suspension to a rearward position, due to the tapered side rails or web area.

Task B1.2   Determine additional auxiliary axle location(s) and install.

Auxiliary axle locations must be determined by an engineer or a qualified individual. A complete and thorough weight distribution must be performed prior to the installation of any load bearing axle. The addition of an auxiliary axle may also require a new brake certification to comply with FMVSS 105 or 121. Brake timing tests for full air brake trucks may be required. Once the engineering side of the installation has been completed, and drawings are made available to the installation technician, they must adhere to those drawings with no deviation, as per Task A7 (layout and drill mounting holes).

Task B1.3   Install additional suspension components including leaf springs, air bags, stabilizer bars (torsion bars), stop blocks, tanks, and valves/controls.

The installation of any additional suspension components must be done with the utmost care and caution. Supplemental suspension components can affect the ride and handling of the completed vehicle. It should be noted that a check of the completed vehicle's chassis and suspension alignment must be completed prior to the delivery of the truck to the end user. Stabilizer and torsion bars are used for multiple purposes. They can be installed to improve the stability of a completed vehicle with a high vertical center of gravity. Or, they can be installed with the purpose of stabilizing the truck when it is stationary in applications where aerial devices are to be used. For example, a "bucket" truck or an aircraft refueling hydrant service truck.

Tanks and control valves are generally used when stationary pneumatic equipment has been added to the truck chassis. This could include additional air suspension auxiliary axles, pneumatically operated throttle controls, and others. The installation of air tanks requires that compressor capacity be examined. Also, the so called "wet" tank and primary tank must be sized for adequate volume. It is also a federally regulated addition that no primary air tank can be connected to an auxiliary air tank without a safety check valve installed between them. Even the addition of an air ride seat, drivers or passengers, must also be checked for compliance with all regulations.

Adding additional pneumatic systems will also require the addition of DOT rated air lines. Most modern air lines are synflex, a plastic material that is mostly impervious to contamination. The routing of these lines and the installation of proper hangers and fittings is an important task.

Task B1.4   Check driveshaft clearances to moving and stationary components under operating condition.

Driveshafts and universal joints are installed by the original equipment manufacturer to very exacting operating angles. Generally, the operating angles of driveshafts and universal joints are available in the chassis manufacturers Body Builder's Manuals. Also, the minimum clearance to the driveline is published. However, a technician must inspect clearance to any component that has been installed on the chassis/cab. This includes power take off units and their related component, pumps, shafts, drive pulleys, hoses, wiring harnesses, all fluid lines, air lines, stabilizer bars, etc. These clearances must be checked in their maximum travel positions, in other words, under both full chassis deflection along with full rebound.

Task B1.5 **Verify proper routing and support of air lines.**

If any pneumatic equipment is added to the chassis, the air lines have to be routed in a safe and effective manner. This means that a technician must comply with the requirements of the chassis Builder's specifications published Body Builder's Manual. On trucks equipped with the current EPA and CARB exhaust emission engines, especially those with diesel engines, clearance around high heat generating equipment such as the Diesel Particulate Filter and catalytic converters is critical. These components may reach temperatures that rise significantly above 700 degrees Fahrenheit. A common practice is to maintain at least 2 to 3 inches clearance from these components. Be particularly aware of bends, kinking, buckling, and stress on all lines and fittings. In addition, be aware of sharp objects that could damage lines and fittings.

Task B1.6 **Attach axle to suspension seats, as necessary.**

In general, axles are attached to suspensions seats using "U" bolts. At the point of contact between the axle and the suspension seat, or spring seats, a wedge shaped seat is installed. Some manufacturers choose to weld this spring seat to the axle housing itself, while others trap the wedge between the corresponding parts. The purpose of the wedge is to set the chassis manufacturer desired rear differential pinion angle. These mounting systems can be complex on both air suspension systems and tandem drive axle systems. The technician is required to insure that "U" bolts being installed are new and of the correct grade, generally grade eight materials. Also, it is imperative that these "U" bolts are torqued to a proper specification to be determined by the vehicle manufacturer. Re-inspection after a period of miles drive should include re-torquing the "U" bolts and an overall inspection of all related parts.

Task B2 Inspection and Repair (2 Questions)

Note: Tasks 1 thorough 10 apply to used chassis components and should be accomplished as necessary.

Task B2.1 **Inspect and replace front axles, "U" bolts, and nuts.**

Virtually all commercial heavy-duty vehicles on the road have a solid front axle assembly that contains king pins on each side that allow the front wheels to turn and thus steer. The axles are connected to the vehicle chassis by springs, "U" bolts, spring shackles, and mounting hardware. It is possible for an axle to be bent and not visible to the eye. Carefully measure the location of the front axle (set back) with a tape measure and use a protractor to check for straightness. Compare all measurements for equality on each side of the vehicle to be certain of correct front end geometry and alignment.

Task B2.2 **Inspect, service, adjust, or replace king pin, steering knuckle bushings, locks, bearings, seals, and covers.**

When inspecting a king pin assembly, the technician will use a dial indicator to establish vertical and horizontal play in the steering knuckle by pressing up and down and back and forth on the assembly. Excessive play indicates worn king pins and possibly worn bushings. If the king pin passes inspection, a service procedure requiring adding grease to the assembly should be completed. Grease the king pins until grease squeezes out of the pivot bearing, and continues to do so until the grease seepage is clean.

When replacing the king pins, remove the upper and lower knuckle caps to access the king pins. Before pressing out the king pin, a technician must remove the lock bolt and drive out the draw key to remove the king pin. The draw key keeps the king pin aligned. King pin bushings are the same on either side. Add shims as necessary to achieve desired clearance. A loose king pin or king pin bearing could cause a shimmy with slight vibrations. Grease the new king pin after installation.

Task B2.3 **Inspect, service, and replace shock absorbers, bushings, brackets, and mounts.**

Shock absorbers dampen or control spring action from jounce and rebound, reduce body sway, and improve directional stability and driver comfort. Shock absorbers can be of the single or double action type. Worn-out shock absorbers present a dangerous driving condition by allowing the vehicle to bounce to the point where stability is compromised. Even vehicles with air suspension system require shock absorbers. Only in very rare situations would a technician replace a shock absorber on only one side of a suspension system, front or rear. Inspect shock absorbers with one end disconnected for resistance, bent piston rods, bushing failure, external damage, and visual leakage. Bear in mind that not all trucks require rear shock absorbers, while front shock absorbers are the norm.

Task B2.4 **Inspect, repair, or replace leaf springs, center bolts, clips, eye bolts and bushings, shackles, slippers, insulators, brackets, and mounts.**

Leaf springs can be mounted either above or below the rear suspension leaves. Mounting them below the axle is the norm, as this will lower the center of gravity of the truck and improve stability. Springs can hide many problems. Be sure to inspect the springs very carefully, especially those that have operated in the snow belt where heavy amounts of road de-icer has been used. Rusted springs may be cracked or completely broken and the crack may be hard to detect without a very close inspection. It is possible that a spring center bolt may be broken, and this too may be very difficult to see as well. Loose spring shackles will not break the spring center bolt. A broken leaf spring center bolt can cause rear axle shift leading to premature toe-like tire wear and steering pull.

Task B2.5 **Inspect, adjust, or replace torque arms, bushings, and mounts.**

Torsion bars are the same length but are usually not interchangeable left to right. As long as a torsion bar is not cracked or broken, it can be adjusted at the crank assembly to level the vehicle. A technician should never heat or bend torsion or torque arms. Premature suspension bushing wear will cause noisy operation, directional instability, and excessive wear to adjacent suspension components. A broken leaf spring can cause an off-level vehicle attitude. A bushing not relaxed during assembly can cause a binding condition, leaving a vehicle off-level.

Task B2.6 **Inspect, adjust, or replace axle aligning devices such as radius rods, track bars, stabilizer bars and bushings, mounts, shims, and cams.**

Tandem axle suspensions often have four multi-leaf springs and four torque rods. Between the front and rear springs on each side of the tractor, the springs ride on an equalizer pivot mounted on a sleeve in the equalizer bracket. Torque rods may be substituted for rear shock absorbers. Four to six torque rods used on spring suspensions are also used for suspension alignments. Rear axle alignment adjustment on a spring suspension with torque rods can be made thorough lower adjustable torque rods. Some rear suspensions use shims between the torque rod front and spring hanger bracket to align the rear axle. Other rear suspensions use an eccentric bushing at the torque leaf for alignment adjustment. Torque rods also provide braking force absorption and driveline angle adjustment.

Task B2.7 **Inspect or replace walking beams, center (cross) tube, bushings, mounts, load pads, and caps.**

Equalizing beam suspension lowers the center of gravity of the axle load. The two types of equalizing beam suspensions are leaf spring type and rubber load cushion type. Tandem rear axle suspension systems have equalizer beams on each side of the suspension. Bushings in the equalizer beams are attached to the rear axle housings and a cross tube is mounted between the two equalizer beams. A technician can service walking (equalizing) beam center and end bushings on the vehicle using a special OEM tool. Multi-leaf springs are mounted on saddles above the equalizer beams.

Some rear tandem axle suspensions have an inverted parabolic tapered leaf spring mounted on a cradle that pivots in a saddle bracket. Rubber springs are mounted between the ends of the springs and saddles on the rear axle housings. Four upper torque rods are mounted between the rear axle housing and the tractor frame, and four lower torque rods are positioned between the rear axle housing and the saddle brackets.

Task B2.8 **Inspect, test, adjust, repair, or replace air suspension pressure regulator and height control valves, lines, hoses, and fitting.**

Prior to beginning any inspection of air suspension components, make sure adequate air pressure is being supplied by the engine driven compressor mounted on the truck's engine. Make sure that recovery time is in compliance with regulations as well. Insufficient air pressure in the base system will reduce the operational efficiencies of all pneumatic systems on the truck, including the brakes.

Perform a visual inspection of the pneumatic system. Checks are to be performed for incomplete, binding, or improperly adjusted linkages; valves mounted in the wrong position; air lines leaking, frayed, cracked, chafing, or improperly positioned; or anything that could lead to future failure.

Note: The most effective means of locating small leaks is a soap-and-water solution sprayed onto all air system components; observe bubbles as an indicator of the leak source.

Task B2.9 **Inspect, test, repair, or replace air bags, springs, mounting plates, suspensions arms, and bushings.**

A visual inspection will determine if the air springs and related mounting hardware are mounted in proper position so as to facilitate proper movement and rigidity. Suspension arms must be checked for bushing wear and movement. Check manufacturer's specifications for each individual system type.

Air springs (bags) should be checked for possible damage, such as cuts or exposed cords. Replacement of air bags is recommended as the cost of the new unit is economical and repair is not suggested.

Task B2.10 **Measure ride height; determine needed adjustments or repairs.**

Always address the manufacturer's service manual specifications to determine:

- If a truck with air suspension includes a dump valve, used to lower the loading height of the truck and improve loading stability when a forklift truck is used, ensure that no interference can take place with tires, and other components.
- The type or model of the suspension, i.e., spring or air type.
- The load rating of the suspension. This is not the GAWR (gross axle weight rating).
- The point(s) of measurement. This varies by the manufacturer and is not the same on all trucks.
- System air pressure and supply (if air type).
- How measurement will be performed, i.e., flat floor, load or no load or trailer attached, brakes applied. Adjust as necessary to specifications.
- What tools will be needed to perform the inspection.

Note: Always perform a height setting on an air spring system with proper air pressure and supply—air dump valve closed and the system air supply valve applying air (air going into the system)—for accurate results. Perform a road test and reconfirm ride height. Improper suspension ride height will cause driveline vibrations, accelerated U-joint wear and changes in geometry to the center of gravity (load), steering, and braking systems.

C. Driveshaft Preparation (5 Questions)

Task C1 **Layout driveshaft; determine new driveshaft length.**

In cases where wheelbase modifications have taken place, or additional equipment has been added to the driveline of the vehicle, such as auxiliary transmissions and power take off units, driveshaft's will require modification by lengthening or shortening. In some cases more than one driveshaft will be involved, and quite possibly a center support bearing as well. Layout of driveshaft for this procedure will require that the technician has a good knowledge of the mathematical equations of driveshaft angularity and phasing.

The technician will need to first determine the overall layout of the driveline. How many shafts will be used? How long will they be? In order to determine driveshaft length, first measure the distance from the drive unit to the driven unit. In some cases, on a single shaft application, this is the distance from the transmission output shaft yoke to the rear differential input shaft, (pinion shaft yoke). In cases where the number of shafts required is more than a single unit, the technician will have to determine the location of a single or double center support bearing and carrier assembly.

It is absolutely imperative that in making this decision the technician is completely familiar with the truck transmission that is being used, and knows the transmission gear ratios, particularly in higher gears and especially in the final gear. Many modern transmissions utilize overdrive ratios in one or more top gear ranges. This means that driveshaft rotational speeds will be faster than engine crankshaft speed. This situation will require shorter driveshaft's than what is considered "normal." Also, due to this situation, the wall thickness of the driveshaft material will be higher than "normal." While 60 inches is considered the longest driveshaft that a truck should be equipped with, a shorter shaft may be required in an application using a high speed engine and an overdrive transmission. Overdrive gearing can be a much as 0.062, which means that while the engine crankshaft speed may be 3000 RPM, the driveshaft may be spinning at 4100 RPM. Where a 0.060 wall thickness may have been considered "normal," a 0.090 or greater wall thickness may actually be required.

If the technician is modifying the original equipment driveshaft that was installed on the truck from the factory, they will need to know what wall thickness was selected. Also, they should know what the maximum length of an original shaft was when installed by the chassis manufacturer. This information can be found in the Body Builder's Manual for each chassis. This manual will also provide what the transmission output shaft (crankshaft) angle and what the differential pinion angle is as well. The technician will have to calculate, based on all of this information, the length, tube thickness, and angles of each driveshaft. It should be noted, that after reconstruction of driveshaft, the complete assembly must be balanced by a professional driveline shop.

Task C2   Inspect driveshaft for proper phasing.

The first commandment of driveshaft disassembly is to scribe onto mating parts "witness" marks. Witness marks should be scribed on slip joint yokes and accompanying driveshaft, transmission yokes and flanges, pinion yokes and flanges and all other mating components. This will help in the reassembly process and insure proper phasing.

Phasing is described as the alignment of single Cardan joints on opposite ends of the driveshaft. On a two driveshaft driveline, it is the alignment of all single Cardan joints in the complete assembly, including the transmission yoke and the pinion yoke. If any of these components are out of phase, torsional vibrations will occur at any speed, and can cause broken driveshaft welds, premature spline wear at the slip joint, and loosening nuts and bolts.

There is no reason that a driveshaft should ever wear out. Failures in new driveshafts or reconstructed driveshafts will occur almost immediately if the driveshaft is assembled out of phase.

Task C3   Install driveshaft; measure and adjust operating angles (loaded and unloaded where applicable).

In order to check operating angles in the driveline, a technician will need an angle finder or protractor, and a level shop floor. The level shop floor is actually optional, because the technician is only interested in finding the difference in angles between components, not the actual angle itself.

Task C4   Lubricate universal joints and splines.

Greasing the universal joints is a relatively easy procedure, but there are a few precautions that a technician must be aware of. Use grease that is specified for universal joints, usually chassis lubrication grease or wheel bearing grease; always confirm with the manufacturer's recommendations.

If the technician is greasing a new universal joint, make sure that grease is added before excess grease starts to discharge from the universal seals or the dust cap. It is important that adequate grease is installed but over-greasing can damage seal integrity.

If the technician is greasing a used universal joint, be aware of water and rust coming out of the joint seals. If excessive rust or water is present, the technician may want to further inspect the universal joint for wear and fatigue.

Task C5   Inspect, service, or replace driveshaft, slip joints, yokes, drive flanges, universal joints, and retaining hardware; properly phase yokes.

A properly fitted driveshaft should not normally wear-out unless the damage is from an external source. However, driveshaft components such as slip joints and universal joints can and do fail most often due to the lack of good maintenance, and carelessness. The installation of an improperly phased driveshaft will become noticeable immediately after installation, as a severe torsional vibration will exist in the driveline. Properly phased yokes are, in simple terms, when the yoke at both ends of the driveshaft are lined up. Prior to disassembly of a driveshaft that contains a slip joint, the driveshaft and the yoke must be marked for reassembly.

Task C6   Inspect, repair, and replace driveshaft center support bearings and mounts.

Center support bearings are used in driveshaft systems where two or more shafts are used. Center support bearings have their own chassis mounting components, usually in the form of a chassis cross member. The dimension of the center support bearing from the manufacturer described location is very important on trucks that have had wheelbase modifications. The center support bearing will be mounted 90 degrees to the driveshaft. Generally, center support bearings are mounted in rubber surrounds and this is a potential trouble spot should the rubber become hard and cracked. Be sure to lubricate center support bearings per the manufacturers' recommendation. Some center hanger bearings are sealed components and cannot be greased.

D. Body and Equipment (24 Questions)

Task D1   Preparation (8 Questions)

Task D.1.1   Verify body/equipment mounting and location on vehicle.

A technician will receive mounting arrangements and instructions from the engineering department and supervisor. It is their job to validate this information. Look carefully at all drawings and instructions. Look at the chassis itself. Perform a walk-a-round of the chassis. Evaluate wire and hose routing, switch locations, cable locations, and equipment locations.

Be sure to validate the location and clearance around power take off units, pumps, and potential interference with frame flanges, frame webs, and cross members. Check for interference with chassis wiring harnesses, piping, hoses, and mounting hardware.

Study the chassis electrical system as well. Look for battery mounting brackets and hardware, electrical grounds, chassis mounted circuit breaker boxes, and electronic control units. Look for "gang" grounds as well where multiple components may be grounded at a single location.

If the truck is equipped with a multiplexed lighting wiring system, become familiar with that system by studying the chassis manufacturers Body Builder's Manual.

Look also for chassis manufacturer provided wiring harnesses that have been designed for the integration of body electrical lighting system such as parking lights, brake and turn signal lights, clearance and identification lights, and side marker lights.

Task D1.2   Select and install proper body spacers as necessary.

The proper clearance of the mounted body to chassis components is important and will change from body application to body application. Clearance between the rear tires, springs, wheels, and suspension components is highly important. Chassis manufacturers provide information in their Body Builder's Manual on desired clearance around these parts.

Of special importance is the clearance between the top of the rear tires and the bottom, or underside, of the body in the rear wheel area. A technician may have to adjust this clearance by adding spacers and/or shims. When considering this clearance, the upper rebound limit of the rear axle is the only satisfactory information that the technician will have at their disposal. Some bodies, in order to attain this desired minimum clearance, will have to utilize wheel boxes in their design. When the customer's request is to mount the body at a specific height, either low or high, a technician will be required to calculate the spacer dimensions using the chassis manufacturer information as published in the Body Builder's Manual.

Where multiple body and equipment combinations are used, such as a sleeper and a van body, or a tunnel tool box and a dump body, each component will require separate mounting and thus, each piece of equipment may require separate spacers, shims, and hardware.

Task D.1.3 **Lay out body and equipment mounting holes; select proper drill sizes.**

As previously pointed out in Task A7 of this publication, the nature of this task requires careful review for hole placement. A technician must establish a plan and request specific information from the engineering groups or OEM guidelines. They must plan ahead. A hole placed in a chassis cannot be repaired easily if it is in an incorrect location.

A heavy paper or light cardboard template may need to be drawn and laid out prior to piercing of the chassis. There are many considerations that must be made to the placement of holes in a chassis.

Holes should not overlap, be elongated or oval shaped. The rule of thumb is that adjacent holes should be no closer than 1 ½ inch diameter from each other's edge. Also the technician should not drill holes larger than ¾ inch diameter; this can cause premature failure and expensive repairs.

When available, follow dimensions published in the chassis Body Builder's Manual or a company's engineering standards manual.

If holes are required on a chassis cross member they should be treated like frames and adequate clearance from corners and bends should also be monitored for the same reason as mentioned above; adequate bolt and washer clearance.

The installation of a hole in a frame should be completed with specialized cutting bits, or twist drills. Cutting fixtures will insure that the hole's diameter is true and precise, that the shoulders are square and that there is no deformation in the hole. Magnetic drills are also the preferred equipment for hole drilling. Drills mounted to stands that can be aligned and secured to the frame are also acceptable. And under no circumstances should a cutting torch or plasma cutter be used to install a hole in a truck frame or cross member. The heat can affect the heat treatment of the frame and cause premature failure.

Task D.1.4 **Identify fastener type, grade, diameter, and length.**

Select fasteners based on the job they must do, not just what is on hand. Cap screw and nuts grades are generally considered the prime method of selection. In most cases, when mounting a typical body such as a van or "box" body, a technician will only be required to use a grade five "U" bolt and mounting plate assembly. However, in cases where the load imposed on the cap screw and washer combination, be it in shear or tension, is exceptionally high, grade eight nuts, bolts and washers may be required.

In cases where extreme loads are imposed, a technician may be required to use a Huck fastener. Huck fasteners are very specialized fasteners that are available in a wide variety of designs. They are extremely strong and exhibit similar properties to a rivet, meaning that they are capable of supporting both tension and shear at the same time. For truck applications, they are most commonly used for installing suspension systems and even frame cross members.

Task D.1.5 **Select appropriate cutting/welding tools and equipment; perform cutting and welding procedures.**

The equipment selected to perform cutting and welding operations will vary by the job at hand. For example, cutting of a truck frame, or another component of similar hardness, construction and size, is best completed with a plasma cutter. In cases where cutting is being performed on a very light-duty component, cutting with a hack saw may be acceptable. Between these two extremes lie many variations. Grinding may be acceptable. Using a high speed cut-off tool may be acceptable. In more crude operations, a cutting torch may be acceptable as well.

Welding operations require a great depth of knowledge of equipment and various metals. A technician may chose gas welding, arc welding, MIG and TIG welding as well. For light-duty welding jobs, MIG and TIG will be suitable. For heavy-duty operations, arc is a must. When selecting and using arc welding systems, there are several rules that must be adhered to. The first is to disconnect the battery at the ground cable attaching point. Insure that all electrical components are isolated from the welding current produced. Two, insure that the materials to be welded are properly prepared. Edges to be welded to must be straight and clean. In some welding instances where metal gauge is high, tapering the ends of each piece of steel to be attached. It is important to avoid defects such as deposited metal cracking, toe cracking, blow holes, slag inclusion, undercuts, and poor penetration.

Task D.2 Installation (16 Questions)

Task D.2.1 Fabricate and install sub-frames as required.

The principle purpose of a truck frame is to provide a platform to mount all components to. This includes engines, transmissions, cabs, suspension systems, and truck bodies and operational equipment. Frames are designed by the truck manufacturers to carry a specific load. However, truck frames are not designed to be interactive with operational equipment without suitable load bearing assisting components like sub-frames. Sub-frames are recommended by many truck chassis manufacturers as a way to protect the truck frame and provide additional mounting systems for components. In applications such as dump bodies, cranes, booms, high-lift aircraft ground support bodies, wreckers and car carriers they are generally recommended. There are cautions that must be taken into consideration when building and installing a sub-frame. When installing a sub-frame the front of the sub-frame should extend as far forward to the cab as possible. This will gradually reduce stress on the truck frame itself by spreading the load over a longer span.

Attaching the sub-frame to the truck frame should not normally be done by welding. Suitable attaching hardware such as mounting brackets that are bolted to the truck frame, or "U" bolts are the preferred mounting methods.

Task D.2.2 Install body, and/or equipment, and related components.

The proper method of installation of the body and equipment will vary by the type of body being mounted. For example, a common van body being installed on a straight truck will generally be mounted using fairly common and simple grade five "U" bolts. But even when using this simple system, care must be taken and rules must be followed.

Task D.2.3 Determine layout, install, connect, and test all federally required lighting and reflector systems.

Using the Federal Commercial Vehicle Lighting Guide that is readily available from the National Truck Equipment Association or available in Title 49 of the Federal Register, insure that all lighting requirements are met. This includes clearance and identification lights, side marker lights, turn signal and brake lights, headlights, and parking lights. It should be noted that clearance and identification lights do not have to be mounted on the truck body, but may in fact be mounted on the truck cab itself.

Connection of these components to the truck chassis electrical system itself in a proper way is paramount to insure long operation of such lights without the development of short circuits, open circuits and other such problems. Many manufacturers provide Body Builder's connection circuits on the chassis itself. These circuits may be used for the installation of brake lights, stop and turn signal lights, etc. When the truck chassis is equipped with such wiring connection points, a technician should try to use them. The Body Builder's Manual from the chassis manufacturer will provide the details on such circuits and connector locations.

Many customers are now requesting the application of Light Emitting Diodes (LED) for turn signal, brake, and body marker lights. In most cases, this does not present a problem to the body installer. The exception to the rules is in the turn signal circuit. Turn signal flasher control modules may not recognize the installation of a LED for a turn signal light. The extremely low current draw of an LED can trick the flasher module into signaling that a bulb is burned out by producing a rapid flash rate. This rate violates the federally established flash rate if no change is made to the flasher unit itself. While some truck manufacturers use a readily available standard three prong flasher that can be converted to a LED compatible unit, others use a turn signal control module that must be replaced with a dedicated unit provided by the chassis manufacturer itself. Check the operation of all lights before delivering the truck to the customer.

Task D.2.4 **Install fuel filler.**

Installation of the fuel filler system must be performed in a manner that allows adequate flow of fuel not only in supplying the fuel to the engine, but also in filling the tank at the pump. The use of DOT approved components in the fuels system is of course required.

Fuel caps may be vented in some cases, or non-vented in others, depending on the type of engine in use. Gas engines will also have carbon canisters installed in order to comply with evaporative emission requirements required by EPA and CARB. In the case of vented system, it is required that a vent system is installed and correctly routed so as not to prevent venting, thus making it hard or impossible to refuel. A technician must insure that vent piping is not kinked, loose, improperly secured and will not chafe by rubbing against a fixed component. If a vent pipe is close to a hot component, the pipe should be shielded as well.

It is desirable to locate the fuel filler pipe and cap in an area that is not close to the filler for another liquid that may be required for operation of the completed vehicle. For example, the fuel tank filler for gasoline to power a generator or a transfer tank of some type should not be located directly adjacent to the fuel filler for the trucks diesel fuel tank.

Diesel fuel systems are required to contain lead plugs in the system that will melt in the event of fire and actually allow spillage in light of the fact that diesel will burn, rather than explode. In the event of a roll over, this is a more desirable method to fight a resulting fire. A common misconception is that these plugs must be located on the fuel cap. In fact, they can also be located directly on the fuel tank itself.

All warning labels regarding the fuel system must be placed on the truck in locations that are obvious to the operator. Distinction between the types of fuel to be used must be obvious.

Task D.2.5 **Determine layout and install federally required rear end/Impact protection.**

Federally required under-ride requirements are for the most part, only applicable to the rear of the truck. However, in some selected applications that are vocational in nature, side under-ride may be required. The compliance department within a company must make that decision.

Rear impact protection must comply with Federal Motor Vehicle Regulations 223 and 224. These regulations define the rear most location on the motor vehicle to be a point that falls above a horizontal plane located 22 inches above the ground and below a horizontal axis located 75 inches above the ground when the motor vehicle is stopped on level ground, unloaded, with tires at their proper inflation pressure, and equipment such as lift gates and cargo doors are all in their normal position as though the vehicle were in motion. Tail lights and other non-structural components are not included in this dimensional location.

There are exceptions for motor vehicles where the protective device may interfere with the operation of the motor vehicle. An example would be a dump truck. In this case, mounting the rear protection device at the rear most location of the truck would prevent the operation of the dump body when tilted. It is possible to mount rear protection in such a way to be legal by adjusting the location to a more forward position while maintaining the 30 inch vertical distance above the ground regulation. A technician's job will be to install such protection that meets not only federal requirements, but local requirements as well. Rear under-ride protection may be welded or bolted in place and may be attached in such a way that a body shear plate may be an integral mounting point.

In cases where trucks will be equipped with lift gate that occupy and operate within the rear protection zone, the National Highway and Transportation Safety Administration has provided an exemption.

Task D.2.6 Construct and install appropriate guards and shields.

Shields and guards may be required in certain applications. Shields may be for heat, or debris. Vehicles that must travel off-road may require shields around such items as transmission coolers, PTO shafts, pumps, and other underbody mounted equipment.

In the case of EPA 2007 diesel engines, particular attention must be paid to the Diesel Particulate Filter, and the Diesel Organic Catalyst. On gas engine vehicles, the same is true for Catalytic Converter. In any regard, adequate clearance from high temperature components must be established. Guards should be installed where rubber components such as fuel lines, bushings as used on shock absorbers, stabilizer bars, brake lines, vent lines, electrical harnesses, and the like may be in close proximity to a high heat generating device.

Task D2.7 Connect auxiliary HVAC.

Auxiliary HVAC may be a system used to provide heat, cooling, or both to a specific vocational portion of a truck. For example, trucks with refrigeration bodies will require auxiliary system. These can be in the form of engine driven refrigeration systems or self contained refrigeration systems. Self contained systems present unique weight distribution considerations, especially on light-duty trucks. Weight distribution analysis must be performed to insure that the front axle of a truck is not overloaded by the additional weight.

Yet another application for HVAC is added sleeper units on both straight trucks and tractors. In these cases, the technician may be required to add a separate and distinct HVAC system to the truck. Use care to install the proper refrigerant in all air cooling systems. R134A is the current legal refrigerant that can be used in all systems. Use of R12 and R22 has been banned. R404 may be requested by the customer wishing to use a (truck) engine driven compressor to freeze payload, however, R404 can produce very high head pressures resulting in component damage.

Task D2.8 Furnish and install informational, operational, and safety labels and manuals in appropriate locations.

Insure that all operating manuals are installed in the truck prior to shipping to the customer. Insure that all applicable warning labels are installed as per instructions provided by the individual equipment manufacturer. Warning labels must be placed in the correct location on the truck as well. For example, lift gate warning labels must be located near the lift gate, not near the front of the vehicle.

Instruction labels for PTO's, auxiliary brakes systems, engine idle speed operation, and others should be mounted near the operational switch of the system itself.

If bilingual labels are specified, they must be installed. Be aware of this in particular on trucks going to Canada.

Task D2.9 Verify that proper required certification labels are attached.

Generally, there are three certification labels available for use. The Final Vehicle Certification Label must be affixed to the truck in a conspicuous place within the drivers cab area. Usually this is the door jamb area or somewhere close to that area. This label states that the vehicle has been completed and is in compliance with all required Motor Vehicle Safety Standards. This label will also contain the Gross Vehicle Weight Rating, the Gross Front Axle Weight Rating, the rear Gross Axle Weight Rating, and the Vehicle Identification Number (VIN). Additional labels may also be required. The Modified Vehicle Certification Label is used when modifications have been made after the completion and certification of the vehicle, such as a snow plow added after the truck is shipped to the truck dealer. Also available is the Altered Vehicle Certification Label. This is used when major changes have been made to the vehicle that effects its final application or a change in Gross Vehicle Weight Rating such as the addition of a load carrying axle.

5 Sample Test for Practice

Sample Test

Please note the letter and number in parentheses following each question. They match the task in Section 4 that discusses the relevant subject matter. You may want to refer to the overview using the cross-referencing key to help with questions posing problems for you.

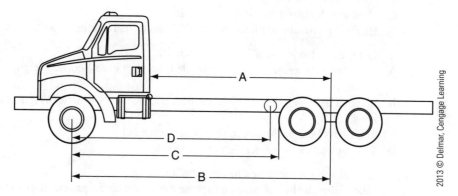

1. The wheelbase dimension of a single axle straight truck is:
   A. The dimension from the back of the cab to the center of the rear wheel
   B. The dimension from the centerline of the front axle to the centerline of the rear axle
   C. The dimension from the front axle centerline to the leading edge of the rear tire
   D. The dimension from the forward rear spring hanger to the front axle (A.1)

2. The usable cab-to-axle is:
   A. Published only as information for the body company
   B. Always 8 inches from the rear window of the cab
   C. The actual location of the front of the body mounting position on the truck
   D. Measured from end of frame to rear most OEM item above frame behind the cab (A.1)

3. Prior to beginning body mounting, a technician should:
   A. Validate all dimensions with drawings and other printed data
   B. Validate all dimensions using the vehicle VIN tag as reference
   C. Insure that the Final Vehicle Manufacturers tag is installed
   D. Measure chassis height behind the cab area (A.4)

4. Tandem axle spreads are:
   A. Always 49 inches
   B. 80 inches or greater
   C. Used in calculating the Bridge Formula
   D. Used in tractor applications for FMVSS length regulations   (A.1)

5. In most states a loaded vehicle is compliant if:
   A. The front axle weight rating is exceeded, but the Gross Vehicle Weight Rating is not
   B. The rear axle weight rating is exceeded, but the Gross Vehicle Weight Rating is not
   C. The Gross Vehicle Weight Rating is not exceeded
   D. The Payload of the vehicle is not exceeded   (A.2)

6. The loaded vehicle frame height could be found by:
   A. Calculating the height based on spring deflection charts provided by the chassis manufacturer
   B. Estimating the defection based on the position of the helper spring pads
   C. Placing a load on the bare chassis and measuring the deflection rate
   D. Using a rule of thumb standard of 0.5 inches deflection per 2000 lbs payload   (A.3)

7. Technician A states that it is not legal to relocate a diesel particulate filter (DPF). Technician B says that it is legal to alter the tail pipe diffuser. Who is correct?
   A. A only
   B. B only
   C. Both A and B
   D. Neither A nor B   (A.5)

8. Per Federal Motor Carrier Safety regulations, a frame mounted battery box must include the following:
   A. A ventilation system
   B. A high current circuit breaker
   C. A cover that is substantial in nature and firmly attached
   D. A structural partition between each battery   (A.5)

9. Technician A states that it is acceptable to weld to a chassis crossmember. Technician B says that welding should not be done on a heat-treated frame. Who is correct?
   A. A only
   B. B only
   C. Both A and B
   D. Neither A nor B   (A.6)

10. What is yield strength?
    A. The point of permanent deflection of a frame under load
    B. The shear strength of a frame
    C. The Resisting Bending Moment (RBM) rating of the frame
    D. The measurement of the volume of steel at a given point on the frame rail   (A.6)

11. Upon inspecting a truck chassis, a crack in the frame flange was noted. Technician A says that the driver may have overloaded the chassis. Technician B says that the frame may be repaired by adding a Fish Plate. Who is correct?
    A. A only
    B. B only
    C. Both A and B
    D. Neither A nor B (A.6)

12. When installing and adjusting a height control valve, which mistake can cause damaging driveline angles?
    A. Improper adjustment
    B. Improper location of the valve
    C. Mismatched parts
    D. None of the above (B2.10)

13. The main purpose of a dump valve on an air suspension system is:
    A. To aid in the stability of a unevenly loaded truck while on the road
    B. To alter the ride when unloading the vehicle while in loading dock
    C. To exhaust all air in the system when the vehicle is parked for long-term storage
    D. To allow the air bags to deflate when chassis is overloaded (B2.10)

14. A pickup truck with a snowplow blade attached and now it appears the front end is too low when the blade is raised. The most likely repair is:
    A. Add ballast in the bed of truck
    B. Install temporary helper springs
    C. Alter the plow attachment bracket design
    D. Perform a wheel alignment and alter ride height (A.4)

15. A used dump truck is leaning to one side. Technician A states that a bad spring pack on the curbside may need to be replaced to reset the height. Technician B states that the frame rails may be twisted and a complete review of suspension is needed. Who is correct?
    A. A only
    B. B only
    C. Both A and B
    D. Neither A nor B (B2.4)

16. After the installation of a variable position fifth wheel, a tractor is coupled to a trailer. The chassis height of the tractor is too low when the trailer is attached. What is the most likely cause?
    A. Air compressor capacity is too low for the application
    B. A leaking air spring
    C. An incorrectly adjusted height control valve
    D. A faulty pressure regulator valve (B2.8)

17. When laying out a new driveline, the driveshaft used should:
    A. Have center support bearing every 60 inches
    B. Be less than 60 inches in overall length to reduce vibration
    C. Defer the OEM and Driveline manufacturer's recommendations
    D. Be designed with the longest shafts available to reduce compound angles (C1)

18. All truck driveshaft wall/material thickness:
    A. Is 0.060 inches
    B. Is 0.120 inches
    C. Will be varied by truck and equipment manufacturers
    D. Is dependent totally on the length of the shaft (C1)

19. Out of phase driveline installation will produce:
    A. A harmonic noise at highway speeds
    B. A torsional vibration
    C. Resonance at low speeds
    D. None of the above (C2)

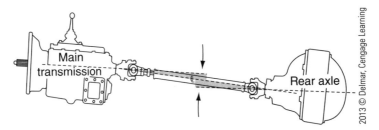

20. Operating angles of universal joints should be:
    A. Within 1 to 3 degrees
    B. Greater than 3 degrees
    C. Between 3 and 4 degrees
    D. The same throughout the driveline of the vehicle (C3)

21. Which of these items would most affect the differential pinion angle under normal driving condition?
    A. The center pin of the rear spring pack is broken
    B. The truck may be over weight on front axle
    C. The truck tires may be under inflated
    D. A worn Torque Arm Bushing (B2.8)

22. Prior to starting to mount an utility body with a corner mounted crane on the chassis, Technician A says that 4-wheel alignment should be performed. Technician B says that the truck should have the spring pack replaced. Who is correct?
    A. A only
    B. B only
    C. Both A and B
    D. Neither A nor B (B2.10)

23. Before working on the trucks electrical system, the first thing the technician should do is:
    A. Check whether all grounds are tight
    B. Disconnect the battery ground cable
    C. Disconnect the air bag system
    D. Disconnect the main fuse box harness connector (A8)

24. A truck in an Altered Stage of assembly was assigned to a technician for Final Stage assembly. The technician should:
    A. Check that all previous work completed was done correctly
    B. Assume that all work was completed correctly
    C. Check to see that the Final Vehicle Certification label has been installed
    D. Check that all instruction sheets, such as the PTO operator's manual, have been placed in the truck cab (B2.9)

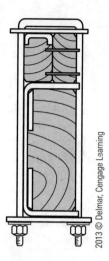

25. A crush block is used to:
    A. Allow the rear under-ride device to deflect when struck from behind
    B. Prevent distortion of the lower frame flange in a "U" bolt location
    C. Strengthen the frame flanges
    D. Prevent the "U" bolts from stretching (D1.2)

26. The "upper rebound limit" of a rear axle is:
    A. The point of maximum spring deflection under load
    B. The maximum allowed travel of the rear axle and all related components
    C. The point at which the rear helper spring bumpers will touch the rear axle housing of the truck
    D. The point of maximum inflation of the rear suspension air bags (D1.2)

27. When installing a wood insulator for a van body installation, the technician should:
    A. Cut the forward end of the insulator at a 90-degree angle
    B. Cut the forward end of the insulator at a 45-degree angle
    C. Taper the front of the body long sill to match the insulator angle
    D. Bolt the insulator to the frame flange (D1.2)

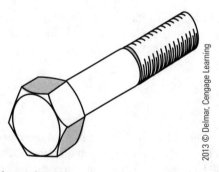

28. Bolt grades are determined by:
    A. Diameter of the bolt
    B. Bolt thread type
    C. Material of the bolt
    D. Bolt flange size (D14)

29. A rivet is used in the assembly of a truck chassis. Technician A says that a comparable bolt can replace it. Technician B says that it can be used to install a crossmember. Who is correct?
    A. A only
    B. B only
    C. Both A and B
    D. Neither A nor B (D1.4)

30. The front of a sub-frame should:
    A. Always be tapered
    B. Be extended as far forward under the body as possible
    C. Be tack-welded to the frame
    D. Be mounted with a "U" bolt mounted as far forward as possible (D2.1)

31. When a sub-frame is constructed that will support two components of the completed vehicle, for example a tunnel toolbox and a dump body:
    A. The sub-frame should be constructed in one piece and therefore support all components of the truck body and equipment
    B. The system shall be constructed as an individual sub-frame for each component
    C. The sub-frame should be constructed for the dump body only and should end just before the toolbox. The toolbox can be mounted directly to the truck frame
    D. The sub-frame should extend from the front of the toolbox to the mounting location of the dump body hydraulic lift cylinder (D2.1)

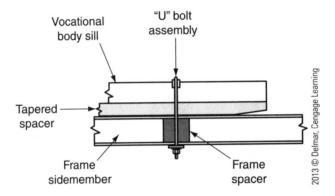

32. The location of the forward "U" bolt on a stake body mounting should be:
    A. As close to the cab as possible
    B. At least 6 inches from the front of the body
    C. In the same location as the first chassis crossmember behind the cab
    D. In the same location as the second body crossmember (D1.3)

33. The "U" bolts supplied with a body mounting kit are too short. Technician A says that the "U" bolts could be replaced. Technician B says that the Shear plates could be installed. Who is correct?
    A. A only
    B. B only
    C. Both A and B
    D. Neither A nor B (D2.2)

34. A platform body with a removable bulkhead requires marker light for FVMSS 108 to meet the frontal identification requirement. Technician A says that the lights should be installed on the platform body. Technician B says that the lights should be a red color to meet the standard. Who is correct?
    A. A only
    B. B only
    C. Both A and B
    D. Neither A nor B (D2.3)

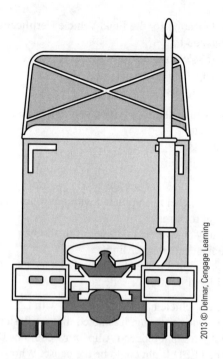

35. The tandem axle tractor in the figure per FMVSS (Federal Motor Vehicle Safety Standards) must be equipped with:
    A. White conspicuity tape at top corners of cab
    B. Red/White conspicuity tape at top corners of cab
    C. Center mounted stop light
    D. Rear back up light   (D2.3)

36. When installing the fuel filler on a utility body, Technician A says that the OEM supplied fuel filler kit should be used. Technician B says that the fill rate tank should be checked prior to delivery. Who is correct?
    A. A only
    B. B only
    C. Both A and B
    D. Neither A nor B   (D2.4)

37. A customer complains that the gas fill rate is too slow after installing a utility body. Technician A says that the OEM supplied fuel filler kit may be the cause. Technician B says that the angle of the fill cup may be the cause. Who is correct?
    A. A only
    B. B only
    C. Both A and B
    D. Neither A nor B   (D2.4)

38. A safety label for a lift gate should be:
    A. Located inside the cab on the sun visor
    B. Located next to the lift gate switch
    C. Shipped with information pack in cab
    D. Thrown away with excess parts   (D2.8)

39. The chassis Incomplete Vehicle Document (IVD) should be:
    A. Attached to the Driver Side door jamb
    B. Kept by the Upfitter of the truck
    C. Discarded by the Upfitter of the truck
    D. Given to the owner of the Upfitted truck   (D2.9)

40. The accepted location for the Final Vehicle Certification Label is:
    A. Under the hood
    B. The driver side door jam
    C. The passenger side door
    D. The driver sun visor (D2.9)

41. An improperly installed fuel system vent may cause:
    A. A fast fill rate
    B. A slow fill rate
    C. Fuel vapors in cab
    D. Overfilling of the tanks (D2.4)

42. A straight truck is being completed with an overall length of 35 feet. This truck will require:
    A. Conspicuity tape at the midpoint of the body, as low as is practical
    B. Intermediate side marker lights
    C. A side turn signal placed in the horizontal center of the body
    D. No additional lighting or reflectors (D2.3)

43. The completed vehicle has been equipped with Light Emitting Diodes turn signal lights. During testing, it is noted that the flash rate is very fast. Technician A says that reduced Amp Draw could be the cause. Technician B says that a bad LED Light could be the cause. Who is correct?
    A. A only
    B. B only
    C. Both A and B
    D. Neither A nor B (D2.3)

44. The air conditioning system is being modified for a van body installation, Technician A says that before refilling the system the gas type will need to be verified. Technician B says to wear safety glasses before venting the gas to the atmosphere. Who is correct?
    A. A only
    B. B only
    C. Both A and B
    D. Neither A nor B (D2.7)

45. When repairing the rear impact bumper on a van body, Technician A says that the Interstate Commerce Commission regulations should be reviewed. Technician B says that the bumper should be welded 30 inches off the ground to improve ground clearance. Who is correct?
    A. A only
    B. B only
    C. Both A and B
    D. Neither A nor B (D2.5)

6 Additional Test Questions for Practice

Additional Test Questions

Please note the letter and number in parentheses following each question. They match the task in Section 4 that discusses the relevant subject matter. You may want to refer to the overview using the cross-referencing key to help with questions posing problems for you.

1. The cab-to-axle dimension of a single axle straight truck is:
    A. Measured from the rear most portion of the cab to the centerline of the rear axle
    B. The cab-to–axle dimension of any truck is used to determine the weight distribution
    C. The published cab-to-axle dimension is always printed on the data tag of the truck
    D. Is measured from the centerline of the front axle to the back of the cab   (A1)

2. The cab-to-tandem dimension is:
    A. The dimension between the cab and the forward rear axle centerline
    B. The dimension between the cab and the rear axle centerline
    C. The dimension from the back of the cab and the centerline between the rear axles on a dual rear axle truck
    D. The dimension from the back of the cab and the front axle centerline   (A1)

3. The bumper to back-of-cab dimension is
    A. The dimension from the rear bumper to the back of the cab.
    B. The dimension is the same as the cab-to-axle dimension on tri-axle trucks.
    C. The dimension from the leading forward edge of the front bumper to the back of the cab.
    D. Only used for weight distribution calculations.   (A1)

4. The addition of an auxiliary axle in a tandem position will:
    A. Always double the payload capacity of a truck
    B. Change the effective wheelbase of a truck and require a weight distribution calculation
    C. Always be in a location that is behind the original drive axle of the truck
    D. Always be in a location that is in front of the original drive axle   (B1.1)

5. The "after frame" dimension:
    A. Is the dimension from the centerline of the rear axle to the end of the frame on a single axle truck
    B. Is required for lift gate applications only
    C. Is fixed and cannot be modified by the chassis Upfitter
    D. Is the dimension from the rear most part of the rear spring rear hanger   (A1)

6. Gross Vehicle Weight Rating is:
    A. The chassis manufactures published maximum weight capacity of the completed vehicle including ALL components and payload
    B. The weight of the vehicle in a completed state with no payload
    C. The capacity of the vehicle when towing another vehicle
    D. Used only for registration purposes   (A2)

7. A Gross Vehicle Weight Rating of 33,001 pounds would be:
    A. Class seven trucks
    B. Class five trucks
    C. Class eight truck
    D. A tractor only   (A2)

8. Gross Vehicle Weight Rating is published:
   A. In the Owner's Manual
   B. In the Incomplete Vehicle Document
   C. On the vehicle data plate only
   D. On the Manufacturer Statement of Origin only (A2)

9. The Gross Vehicle Weight Rating is:
   A. Used in the compliance certification for Federal Motor Vehicle Safety Standard 121, Air Brakes
   B. Affected by the installation of a snowplow
   C. Not applicable for truck tractors
   D. Found on powered units only (A2)

10. The published Front Gross Axle Weight Rating and Rear Gross Axle Weight Rating must be equal to:
    A. The Tare Weight of the complete vehicle
    B. The Gross Combined Weight Rating
    C. The Gross Vehicle Weight Rating
    D. None of the above (A2)

11. The Tire and Wheel data label is:
    A. Listed only in the Body Builder's Manual
    B. Mounted to the cab in a conspicuous location like the doorpost
    C. Not required by law
    D. Found only on trailers (D2.9)

12. The Gross Vehicle Weight Rating and the individual Gross Axle Weight Ratings must be affixed:
    A. To the vehicle via the Final Vehicle Certification Label
    B. To the truck body
    C. To the body bill of sale
    D. To the completed vehicle only when a change has been made to the vehicle such as the addition of an auxiliary axle (D2.9)

13. The Gross Vehicle Weight Rating can be changed:
    A. By the Final Vehicle Manufacturer without authorization by the Incomplete Vehicle Manufacturer
    B. Only with the approval of the Incomplete Vehicle Manufacturer
    C. As long as the individual axle weights ratings are also changed
    D. The GVWR can never be changed (D2.9)

14. The Rear Gross Axle Weight Rating must be changed when a:
    A. Lift gate is installed
    B. Forklift will be used to load the truck
    C. Tow truck is built
    D. None of the above (D2.9)

15. Technician A states that the Incomplete Vehicle Document will define what the vehicle can be certified as. Technician B states that the Final Stage Decal is legal document certifying compliance to Federal Standards. Who is correct?
    A. A only
    B. B only
    C. Both A and B
    D. Neither A nor B (D2.9)

16. Frame height is generally measured:
    A. At the extreme rear of the frame
    B. Right behind the cab
    C. At the rear spring front hanger
    D. Where the manufacture publishes it according to the Body Builder's Manual (A3)

17. Technician A says the body height can affect the tire jounce clearance. Technician B says that the body height can be changed by use of body insulators. Who is correct?
    A. A only
    B. B only
    C. Both A and B
    D. Neither A nor B (A3)

18. Technician A says that most frame widths on North American chassis is 34 inches. Technician B says that most frame heights on North American trucks is 11 inches. Who is correct?
    A. A only
    B. B only
    C. Both A and B
    D. Neither A nor B (A3)

19. The "usable" length of a truck frame is:
    A. Measured from the front of the frame rail to the rear
    B. The available frame to mount equipment
    C. Back of the cab to the end of the frame
    D. An exact measurement supplied by the OEM (A1)

20. Technician A says the frame may need to be cut shorter when mounting a dump body. Technician B says the frame rail could be shortened behind the rear axle with a cutting torch. Who is correct?
    A. A only
    B. B only
    C. Both A and B
    D. Neither A nor B (A6)

21. Technician A says that the overall width of a completed vehicle should be measured at the widest point of the body. Technician B says the overall length of the completed vehicle can affect the weight distribution. Who is correct?
    A. A only
    B. B only
    C. Both A and B
    D. Neither A nor B (A2)

22. The overall length of a straight truck with a single rear axle:
    A. Is legally a maximum of 40 feet
    B. Is governed by state and local ordinances
    C. Is measured from the front of the cab and does not include the bumper
    D. Is no more than 150 percent of the wheelbase (D2.2)

23. Technician A says that before working on a vehicle, a review of the chassis information provided by the OEM should be completed. Technician B says before working on the vehicle check that all OEM systems are operating correctly. Who is correct?
    A. A only
    B. B only
    C. Both A and B
    D. Neither A nor B (A10)

24. A heated dump body is to be installed on diesel engine with a after treatment system. Technician A says the after treatment system cannot be altered without OEM approval. Technician B says the body to cab distance may need to be increased. Who is correct?
    A. A only
    B. B only
    C. Both A and B
    D. Neither A nor B (A4)

25. The exhaust pipe behind a Diesel Particulate Filter can be modified:
    A. Only when the OEM approves the modification
    B. Should not be modified under any circumstances
    C. Only if the engine displacement is over 11 liters
    D. If the truck will be designated for off-road use only (A4)

26. Relocating fuel tanks to the opposite side of the truck is permitted only when:
    A. A Letter of authority has been received from the chassis manufacturer
    B. A lateral center of gravity has been completed
    C. Compliance with all Federal Motor Vehicle Regulations has been determined
    D. The fuel tank is structurally constructed (A4)

27. Technician A says that relocating vehicle air tanks should be avoided. Technician B says that when adding a Tag Axle, an air tank may need to be added. Who is correct?
    A. A only
    B. B only
    C. Both A and B
    D. Neither A nor B (B1.2)

28. Chassis mounted anti-lock brake control modules:
    A. Are illegal to relocate
    B. Can be relocated with extreme care and as a last resort only and must be protected from high heat components such as exhaust system components
    C. Anti-lock brake control modules are never chassis mounted
    D. Will only be found on class eight tractors (A4)

29. The correct tool to use to install holes in truck frame is a:
    A. Oxygen Acetylene cutting Torch
    B. Pneumatic Hand Drill
    C. Magnetic slug cutter
    D. Plasma cutter (A7)

30. Rivets should be removed from a chassis using a:
    A. Twist drill on a magnetic drill
    B. A broach specially designed for rivet removal
    C. A cutting torch
    D. A cut-off grinder (D1.5)

31. Prior to starting any component relocation work on frame-mounted components, the Technician should first:
    A. Drain the fuel tank for safety purposes
    B. Ground the chassis
    C. Disconnect/isolate the batteries
    D. Disconnect the alternator main harness (A8)

32. The Technician needs to repair a cracked cab-mounting bracket to a frame rail. The Technician should:
    A. Weld the bracket to the frame web only
    B. Modify the bracket so it can be bolted on
    C. Add a Fish Plate to reinforce the frame
    D. Add a second mounting bracket to reduce the load on the original (D1.5)

33. The Technician needs to lengthen the fuel lines during a fuel tank relocation. The Technician should choose a line based on which of the following:
    A. Only meet SAE requirements
    B. Only be routed in such a way as to prevent chaffing and exposure to heat
    C. Only be of equal diameter to the original equipment hoses
    D. All of the above (D2.4)

34. A Technician altered the wheelbase on a tandem axle platform body to accept a longer body. When lengthening anti-lock brake wiring harnesses, Technician A says to lengthen the wires and solder joints and seal with waterproof electrical tape. Technician B says that the ABS control module needs to be calibrated using an OBD II interface tool. Who is correct?
    A. A only
    B. B only
    C. Both A and B
    D. Neither A nor B (B1.1)

35. A Technician should never place holes in:
    A. Frame webs
    B. Frame flanges within the wheelbase of the truck
    C. Frame flanges in the after-frame area
    D. Crossmembers (A7)

36. A utility body is being modified to accept a corner-mounted crane. The customer plans on using the vehicle in a severe off duty area. The Technician should mount the crane with:
    A. Vocation specific torque prevailing fasteners
    B. Grade eight mounting bolts, nuts, and washers
    C. Special vibration resistant Huck fasteners
    D. Supplied mounting hardware and directions from the crane supplier (D1.4)

37. Technician A says the RBM (Resistance Bending Moment) of a frame rail can be affected by welding. Technician B says the RBM (Resistance Bending Moment) can be increased by Fish Plating the frame. Who is correct?
    A. A only
    B. B only
    C. Both A and B
    D. Neither A nor B (A6)

38. When shortening a wheelbase:
    A. It is important to keep all crossmembers in their original locations
    B. A Technician must only relocate the rear spring hangers to a more forward position
    C. A Technician must only cut and splice the frame rail to the new shorter dimension
    D. A Technician may slide the rear axle forward or cut, splice and reinforce the frame at the point where it was cut (A6)

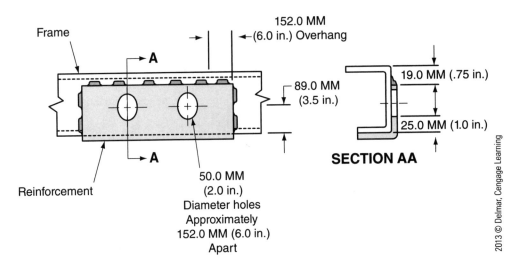

39. Technician A says the picture above shows the proper method to add a "L" section reinforcement to a frame. Technician B says that the welds are spaced to reduce stress concentration on the frame rail. Who is correct?
    A. A only
    B. B only
    C. Both A and B
    D. Neither A nor B (A6)

40. A Technician can extend the wheelbase of a tapered frame truck by:
    A. Cutting and installing a suitable frame extension and reinforcement section in the spliced area
    B. Sliding the drive axle rearward, and adding a reinforcement section to the after-frame
    C. Trucks manufactured with tapered frames cannot be lengthened
    D. Only on trucks with frame yield strength ratings of 50,000 psi or more (A6)

41. Technician A says the Yield Strength is determined by the thickness of the frame. Technician B says altering the wheelbase can affect the Yield Strength of the frame. Who is correct?
    A. A only
    B. B only
    C. Both A and B
    D. Neither A nor B (A6)

42. During a front spring repair, the Technician changed a wedge shaped shim mounted between the axle and the leaf spring on a straight axle truck. Which would be most affected?
    A. Toe
    B. Camber
    C. Caster
    D. Wheel cut angle (B2.2)

43. Technician A says that before installing a frame-mounted crane the RBM (Resistance Bending Moment) of the frame needs to be checked. Technician B says that before installing the crane a Fish Plate needs to be installed to reinforce the frame. Who is correct?
    A. A only
    B. B only
    C. Both A and B
    D. Neither A nor B (D2.2)

44. Technician A says that Aluminum frames are used by some manufacturers to reduce the weight of the body. Technician B says that Aluminum frames are more flexible and most used in off-road applications. Who is correct?
    A. A only
    B. B only
    C. Both A and B
    D. Neither A nor B   (D1.1)

45. When mounting a body on a used chassis, it is imperative to inspect the frame for:
    A. Cracks in the web or flanges only
    B. Cracks in the crossmembers only
    C. Cracks in between holes in the frame only
    D. All of the above   (A9)

46. A common crack, one that extends from one frame hole to another:
    A. Cannot be repaired
    B. Can be repaired by installing hardened nuts and bolts, along with hardened washers that are larger than the twice the diameter of the holes
    C. Welding the crack using proper welding procedures
    D. Installing a fish plate on each side of the frame flange   (A9)

47. A customer has requested a larger under body toolbox to be installed on the curbside of the vehicle. It will be located next to the exhaust catalyst. The box can be placed:
    A. Within 1 inch of the Diesel Particulate Filter (DPF) and related hardware
    B. Within 3 inches of the DPF and related hardware
    C. At a distance recommended by the chassis manufacturer and published in the Body Builder's Manual
    D. Wherever it is the most convenient and cost effective   (D2.2)

48. Technician A says to properly weld a frame crack, the crack itself must be "V" ground with a grinder. Technician B says that to properly repair the frame crack the welding rod should have a tensile strength rating exceeding the frame rail. Who is correct?
    A. A only
    B. B only
    C. Both A and B
    D. Neither A nor B   (A6)

49. To properly lay out a new hole pattern on a truck frame, a Technician should:
    A. First, create a pattern on a suitable material, then measure twice, pierce once
    B. Use a protractor to be sure the frame is level when the template is made
    C. Use the part/component that is to be relocated/mounted to establish a pattern directly to the frame itself
    D. Heat the frame to soften it and make drilling easier and therefore cleaner   (A7)

50. Technician A says that when cutting holes in a frame it's best not to exceed 1 inch diameter. Technician B says that when cutting a hole in the frame rail it's best to stay in the center of the frame web as possible. Who is correct?
    A. A only
    B. B only
    C. Both A and B
    D. Neither A nor B   (A7)

51. Which frame material will be more difficult to pierce?
    A. 44,000 psi Steel (Yield Strength)
    B. 56,000 psi Steel (Tensile Strength)
    C. Cold rolled steel
    D. 110,000 psi Steel (Yield Strength)   (A7)

52. What is the purpose of disconnecting the batteries of a truck prior to welding on it?
    A. To protect the trucks electrical system, including electronic control modules, from damage
    B. Just to eliminate the potential of spark ignited fires in the work place
    C. To protect the Technician from physical harm
    D. All of the above (A8)

53. Technician A states it is recommended by all chassis manufacturer that all harnesses are unplugged from every Electronic Control Module prior to welding to a truck chassis. Technician B says that removing the battery ground wire is adequate. Who is correct?
    A. A only
    B. B only
    C. Both A and B
    D. Neither A nor B (A8)

54. Technician A states that when welding to a chassis that is suspended above the ground, jack stands should be used to support the vehicle. Technician B says that when the chassis is off the ground, the battery should be disconnected. Who is correct?
    A. A only
    B. B only
    C. Both A and B
    D. Neither A nor B (D1.5)

55. Which component is most likely to be damaged by welding to a frame when the battery has not been disconnected?
    A. The starter motor
    B. The Electronic Control Module
    C. The instrument panel
    D. The truck batteries (A8)

56. A dimpled top flange on a new truck was most likely caused by:
    A. Accidental damage during manufacturing
    B. The improper loading and securing of the chassis during the shipping process, in particular, the use of a piggy back trolley mechanism for stacking and carrying chassis on each other
    C. The use of chains wrapped around the frame flange and connected to a trailer to secure a truck chassis during shipping
    D. Both B and C above (A9)

57. It is important to check a frame for squareness. This can be done by:
    A. Using a tape measure
    B. Using a line of site measurement
    C. Using a string
    D. Using a straight edge (A9)

58. On a used truck, elongated holes in a frame where the shear plate was located is most likely the result of:
    A. The holes were drilled too large in the shear plate
    B. The fastener was under torqued
    C. Frame rail steel yield strength that was not high enough for the application
    D. Corrosion from improper application of undercoating (A9)

59. A cracked frame crossmember was found during an inspection. The Technician should:
    A. Replace the member with a new part
    B. Weld the member at the crack
    C. Remove the member and fabricate one with heavier material
    D. Weld the member and install Huck fasteners (A10)

60. A good indication of a bent truck frame is:
    A. When the wheelbase dimension of each side of the truck is more than one quarter of an inch different
    B. When the frame rail has a crack in a web section
    C. When the outside edges of the front tires are worn
    D. When a leaf spring center bolt has broken more than once (A9)

61. If a Technician suspects that a frame is "bowed" to the right or the left, a quick device that can be used to make a preliminary inspection is:
    A. A string long enough to extend from the forward most accessible portion of the frame to the rear, checking the gap between the string and the frame web along the way
    B. A straight edge at least 36 inches long that a Technician can use to check sections of the frame
    C. A tape measure measuring the frame width every 2 feet of frame length
    D. A visual inspection is usually enough to make a sound determination (A9)

62. When inspecting a used truck chassis, a Technician finds a location on the bottom flange of the frame, near the transmission Power Take Off opening that has been squarely ground. This is generally a sign that:
    A. Originally a PTO was too large to install and the lower frame flange was ground to accommodate the PTO and provide clearance. This must be repaired prior to completing the body installation
    B. The frame has been repaired at some point in time
    C. This was a factory installed clearance point and no corrective action is required
    D. The frame cannot be repaired and a replacement frame job is in order (A9)

63. Technician A states that heat-treated truck frame that is bent cannot be repaired. Technician B states that a thorough inspection of the rest of the suspension should be performed. Who is correct?
    A. A only
    B. B only
    C. Both A and B
    D. Neither A nor B (A10)

64. The installation of a "pusher" axle will:
    A. Have the effect of lengthening the wheelbase
    B. Have no effect on the wheelbase calculation for weight distribution analysis
    C. Have the effect of shortening the wheelbase
    D. Have no effect on the payload of the truck (B1.2)

65. When properly installed, the insulator used between the body and the truck chassis:
    A. Is used only to act as a sacrificial material to prevent damage to the truck frame flange
    B. Is used only to provide a flexible surface allowing the body and frame to "move" on one another
    C. Is used only to provide improved ride quality
    D. All of the above (D1.2)

66. A crossmember is normally used in a truck suspension to:
    A. Provide torsional rigidity to a frame rail
    B. Provide a mounting location for body equipment such as engine driven compressors
    C. Provide supported mounting locations for fuel tanks and battery boxes
    D. Provide static flexibility to a truck chassis (A10)

67. After lengthening a wheelbase, another crossmemeber is required. The Technician should always attach it to the:
    A. Frame flanges
    B. Frame web
    C. Center Support Bearing
    D. All of the above (A10)

68. What style of crossmember is most common on a medium-duty truck?
    A. Tubular
    B. Alligator
    C. I beam
    D. Cross beam (A10)

69. The rear spring hanger brackets have a crack in the casting, Technician A says that bracket can be brazed instead of replacing it. Technician B says the leaf spring pivot bolt must be replaced. Who is correct?
    A. A only
    B. B only
    C. Both A and B
    D. Neither A nor B (A10)

70. During a suspension inspection, it is noticed that the front shock absorbers have oil leaking from the seals. Technician A says that the shock absorbers need to be replaced. Technician B says that the leaking air bag can be the cause. Who is correct?
    A. A only
    B. B only
    C. Both A and B
    D. Neither A nor B (A10)

71. Which of the following conditions can result in a vehicle standing in an off-level condition when loaded?
    A. A broken helper spring
    B. A spring that shows signs of rust, but is not broken
    C. An improperly lubricated rear suspension bushing
    D. A broken shock absorber (B2.4)

72. The rear spring shackles:
    A. Allow for changes in the length of a leaf spring while it is in an oscillating motion
    B. Allow control of the horizontal plain of the spring
    C. Stabilize the vehicle when under load
    D. Are only used on the helper spring mounting system (B2.4)

73. Double acting shock absorbers help reduce:
    A. Bounce in only the upper rebound event
    B. Bounce in both the up and down motion of a chassis
    C. Sideslip of the rear axle
    D. Pinion angle changes during acceleration (B2.3)

74. When greasing a front axle king pin, grease comes out of pivot bearing. What is the cause?
    A. There is a leaking seal in the king pin housing
    B. The king pin bushings are worn
    C. The grease is incorrect. Higher viscosity grease is required
    D. There is no failure, as this show that adequate penetration of grease has been accomplished (B2.3)

75. Technician A says the purpose of a torque arm on a suspension system is to help to retain axle alignment. Technician B says it helps control axle torque. Who is correct?
    A. A only
    B. B only
    C. Both A and B
    D. Neither A nor B (B2.5)

76. While performing a wheel alignment on a straight axle suspension, the camber reading is too far out of tolerance. Technician A says that the spring seat needs a shim installed to bring it within tolerance. Technician B says that excessive camber can cause wheel wear. Who is correct?
    A. A only
    B. B only
    C. Both A and B
    D. Neither A nor B (B2.6)

77. Driveshaft length generally includes:
    A. Only the tube assembly
    B. The shaft and yoke ends
    C. The total length of all driveshafts in combination
    D. There is no set method to describe driveshaft length (C1)

78. Phasing of a drive shaft is:
    A. The perfect alignment of the yoke ends on each end of the driveshaft
    B. When two yokes are aligned at 90 degrees of each other
    C. When the complete driveshaft system is balanced as one complete unit
    D. When the crankshaft inclination angle and the rear pinion angle are exactly the same (C2)

79. A customer complains of a vibration when truck is at highway speeds. Technician A says that driveline balance could be the cause. Technician B says that the working angles of the driveline could be the issue. Who is correct?
    A. A only
    B. B only
    C. Both A and B
    D. Neither A nor B (C5)

80. Technician A states that when the operating angle at the drive end of a driveshaft is too large, a secondary couple vibration will occur. Technician B states that large operating angles can cause universal joint failures. Who is correct?
    A. A only
    B. B only
    C. Both A and B
    D. Neither A nor B (C3)

81. The crankshaft angle of the vehicle is:
    A. Always the same as the rear differential pinion angle
    B. Never more than 5 degrees from level
    C. The same angle as the transmission main shaft angle
    D. Variable and will change as torque is created (C3)

82. If a universal joint cross shaft is cracked through the lubrication fitting hole, the problem was Most-Likely caused by:
    A. Universal joint normal wear over time and mileage
    B. Incorrect orientation of the grease fitting when assembling the shaft
    C. A shock load
    D. Incorrect operating angles (C4)

83. During the inspection of a universal joint it exhibits some looseness. Technician A says a lack of lubrication could be the cause. Technician B says that the worn needles bearing should be replaced. Who is correct?
    A. A only
    B. B only
    C. Both A and B
    D. Neither A nor B (C4)

84. Driveshaft twist is usually caused by:
    A. Shock loads
    B. Constant starts in second or third gear
    C. Installing yokes out of phase
    D. A broken universal joint (C5)

85. Checking drive shaft straightness should be done:
    A. With a straight edge
    B. By rolling the driveshaft on a flat surface
    C. Place the driveshaft in a fixture and checking the shaft with a dial indicator
    D. Only during the balancing process (C5)

86. Critical speed failures are due to which of the following combinations:
    A. A driveshaft that is too long and a shaft wall thickness that is too thin
    B. A driveshaft that is too short, in combination with an overdrive high gear transmission
    C. A driveshaft that was assembled out of phase and with incorrect operating angles at the universal joints
    D. A rear pinion angle that is too severe, and weak springs that allow the differential to rotate when torque is applied (C5)

87. Crush blocks in the frame rail should be installed:
    A. Where "U" bolts are located
    B. At 18 inch intervals along frame rail
    C. On frame under the van body
    D. At each frame crossmember (D2.2)

88. The best material to use for crush block construction is:
    A. Oak
    B. Pine
    C. Steel
    D. Polypropylene (D1.1)

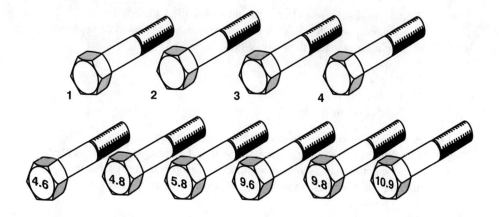

1. Grade 2 (GM 200-M)
2. Grade 5 (GM 200-M)
3. Grade 7 (GM 200-M)
4. Grade 8 (GM 200-M)
5. Manufacturer's Identification
6. Nut Strength Identification
7. Identification Marks (Posidriv Screw Head)

89. A bolt head in the picture shown is a:
    A. Grade 5 bolt
    B. Metric 8.8 class bolt
    C. Grade 8 bolt
    D. Grade 2 bolts   (D1.4)

90. All of these materials would make a good drilling template EXCEPT for:
    A. Thin gauge sheet metal
    B. Heavy gauge cardboard
    C. Sheet stock of magnetized soft, durable and pliable material similar to that used for certain refrigerator magnets
    D. A one 8 inch sheet steel plate   (D1.3)

91. A Technician will need to fabricate a sub-frame to improve tire jounce clearance. Which material would best?
    A. Box channel steel
    B. "C" channel steel
    C. "I" beam shaped steel
    D. Either Box or "C" Channel   (D2.1)

92. The mounting of the sub-frame to the truck should be done:
    A. By welding the sub-frame to the frame as long as it is not welded to the top flange area
    B. By welding suitable steel brackets/plates to the frame web area and the sub-frame side member
    C. By bolting the sub-frame to the truck frame directly through the top frame flange
    D. By fabricating suitable brackets and bolting them to the truck frame web area   (D2.1)

93. Technician A says a sub-frame should be installed when mounting an aluminum body. Technician B says the sub-frame should be made of aluminum to match the body. Who is correct?
    A. A only
    B. B only
    C. Both A and B
    D. Neither A nor B (D2.1)

94. A Technician will need to alter the after-frame length on a truck with Multiplexed Wiring. Technician A says that if welding procedure is used, then the battery needs to be disconnected. Technician B says that the altering the wheelbase may need a re-flash of the Multiplex System. Who is correct?
    A. A only
    B. B only
    C. Both A and B
    D. Neither A nor B (D1.5)

95. The Federal Motor Vehicle Safety Regulation that dictates lighting requirements is:
    A. FMVSS 209
    B. FMVSS 121
    C. FMVSS 108
    D. FMVSS 111 (D2.3)

96. Technician A says fuel filler and hoses should be routed away from the exhaust. Technician B says that the fuel filler from the chassis manufacturer should not be modified. Who is correct?
    A. A only
    B. B only
    C. Both A and B
    D. Neither A nor B (D2.4)

97. A customer has a request to have tamper resistant fuel filler. Technician A says the fuel filler location may be inside the body as long as it is ventilated. Technician B says that the fill cup could have a lock placed on it. Who is correct?
    A. A only
    B. B only
    C. Both A and B
    D. Neither A nor B (D2.4)

98. Technician A states that when completing the final stage decal the decal should contain the Vehicle Identification Number. Technician B says it must contain the Serial Number of the body. Who is correct?
    A. A only
    B. B only
    C. Both A and B
    D. Neither A nor B (D2.9)

99. The tire and rim label is affixed:
    A. To the vehicle by the chassis manufacturer
    B. To the outside of the body at each wheel location
    C. Is not affixed to the vehicle
    D. Is located on the glove compartment door (D2.9)

100. After a new platform body has been installed, Technician A states that a Final Vehicle Certification Label is installed if it's classified as a used truck. Technician B says that an Altered Vehicle Certification label is installed if it's classified as a completed vehicle prior to the first retail sale. Who is correct?
    A. A only
    B. B only
    C. Both A and B
    D. Neither A nor B

(D2.9)

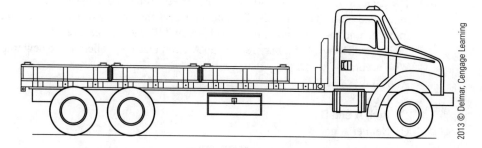

101. The Technician notices the tire clearance in the figure shown may need to be increased. Technician A says that a long sill on the body should be removed and replaced. Technician B says that a sub-frame may be added under the body. Who is correct?
    A. A only
    B. B only
    C. Both A and B
    D. Neither A nor B

(D1.2)

102. The figure above shows the driven axle shaded installed on a van body truck. Technician A says the tag axle is the rear most axle. Technician B says the addition of the tag axle will increase the payload. Who is correct?
    A. A only
    B. B only
    C. Both A and B
    D. Neither A nor B

(B1.2)

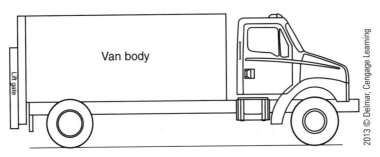

103. The dealer wants the body and liftgate removed from the old chassis and reinstalled onto a new chassis. Technician A says that it may be easier to cut the frame behind the axle and remove the body and liftgate as one assembly. Technician B says that it may require an Altered Decal installed in the new truck. Who is correct?
    A. A only
    B. B only
    C. Both A and B
    D. Neither A nor B

(D2.9)

104. All these items are correct about this decal EXCEPT
    A. The label is affixed to the streetside hood
    B. The decal adhesive must last 10 years
    C. The label is designed to be detached for customer decal
    D. The label includes the Vehicle Identification Number

(D2.8)

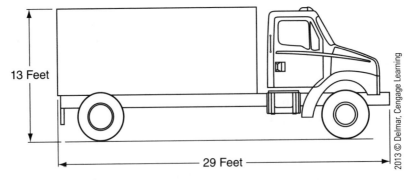

105. Technician A says the truck in the picture shown requires an Amber Light in the center of the chassis. Technician B says that the truck in the picture requires lights at the upper edge of the body corners. Who is correct?
    A. A only
    B. B only
    C. Both A and B
    D. Neither A nor B

(D2.3)

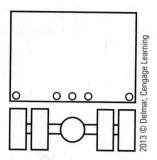

106. The Technician is reviewing the rear of a newly installed van body. Technician A says that if the body under 80 inches wide it does not require reflectors. Technician B says that if the body is lower than 30 inches it does not require a bumper. Who is correct?
    A. A only
    B. B only
    C. Both A and B
    D. Neither A nor B

(D2.5)

107. A center high mounted stop light is required to be installed for all these reasons EXCEPT
    A. The chassis has single rear wheels
    B. The chassis width is less than 80 inches
    C. The chassis is less than 10,000 pounds
    D. The chassis has dual rear wheels

(D2.3)

108. The Technician is installing the proper lights to the rear of a class 8 vehicle. Technician A says that the rear of the vehicle requires an Identification Red Colored Tri-Light. Technician B says that the Identification Lights are to be spaced from center to center 6 to 12 inches apart. Who is correct?
    A. A only
    B. B only
    C. Both A and B
    D. Neither A nor B

(D2.3)

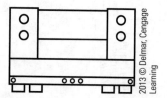

109. The customer has requested a special back up alarm that would use up one of the available pre-cut light holes in the figure above. Technician A says that the back up alarm can replace one of the back up lights. Technician B says that the back up alarm requires a minimum rating of 103 decibels. Who is correct?
    A. A only
    B. B only
    C. Both A and B
    D. Neither A nor B

(D2.3)

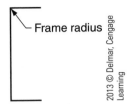

110. The picture shown represents a heat-treated frame rail that will require body shear plates installed. Technician A says the Web area is typically 2 inches away from the frame radius. Technician B says that Web of the frame is generally the best area to drill holes. Who is correct?
    A. A only
    B. B only
    C. Both A and B
    D. Neither A nor B   (D1.2)

111. A utility body with a welding machine mounted in the bed of the truck is leaning 2 inches lower on the Curb Side and the Street Side. Technician A says to add a leaf spring to the Curb Side Spring Pack. Technician B says that the vehicle weight rating will be affected by the added leaf spring. Who is correct?
    A. A only
    B. B only
    C. Both A and B
    D. Neither A nor B   (B1.3)

112. When installing a under deck air compressor, Technician A says to maintain at least 8 inches from the rotating driveshaft. Technician B says that drive shaft safety hoop maybe needed to protect compressor. Who is correct?
    A. A only
    B. B only
    C. Both A and B
    D. Neither A nor B   (B1.4)

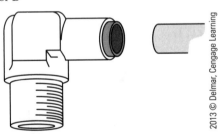

113. Technician A says that fitting below is designed to push into the connector to seal the line. Technician B says that it is important to route the air line so that it cannot pull on the fitting. Who is correct?
    A. A only
    B. B only
    C. Both A and B
    D. Neither A nor B   (B1.5)

114. The rear shock absorber mount is cracked at the rear axle. Technician A says to replace the bracket and follow the manufacturer's replacement procedure. Technician B says to review the payload of the vehicle to prevent future occurrences. Who is correct?
    A. A only
    B. B only
    C. Both A and B
    D. Neither A nor B   (B1.6)

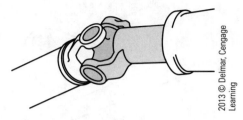

115. Technician A says the slip joint on the driveline alters the length of the driveline. Technician B says the slip joint on the driveline helps seal the transmission case. Who is correct?
    A. A only
    B. B only
    C. Both A and B
    D. Neither A nor B   (C1)

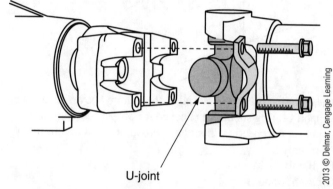

U-joint

116. The U-Joints on the driveline were changed due to bearing wear. Technician A says to check the shaft phasing before torquing the end caps. Technician B says to grease the joint before torquing the end caps. Who is correct?
    A. A only
    B. B only
    C. Both A and B
    D. Neither A nor B   (C2)

117. The U-Joint failed in the driveline. Technician A says the working angle was 3 degrees rear axle. Technician B says the driveline could be too long. Who is correct?
    A. A only
    B. B only
    C. Both A and B
    D. Neither A nor B   (C1)

118. All of these could cause driveline failure EXCEPT
    A. Incorrect U-Joint phasing
    B. 5 degree work angle
    C. Driveline balanced on bench
    D. Over greasing the U-Joint   (C3)

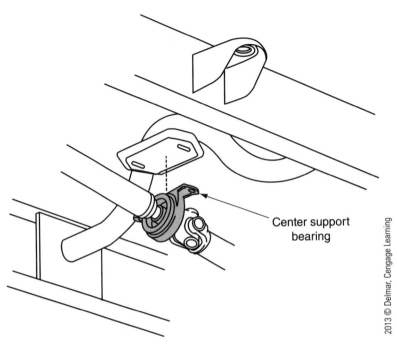

119. The driveline is being serviced in the picture above. Technician A says the Center Support bearing should be checked for proper lubrication. Technician B says the center support bearing should be perpendicular to the frame. Who is correct?
    A. A only
    B. B only
    C. Both A and B
    D. Neither A nor B

(C6)

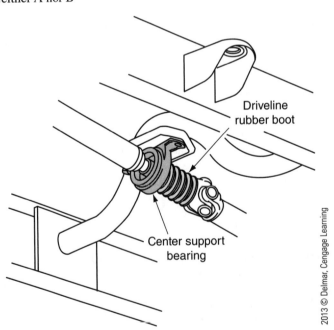

120. The center support bearing in the picture shown makes a growling noise while under load. Technician A says the rubber support for the bearing may have failed. Technician B says the driveline rubber boot may have caused the bearing to fail. Who is correct?
    A. A only
    B. B only
    C. Both A and B
    D. Neither A nor B

(C6)

121. A van body is being installed a new single rear wheel cab chassis. Technician A says to mount the body 3 inches from back of cab. Technician B says "U" bolts must be used to mount the body. Who is correct?
    A. A only
    B. B only
    C. Both A and B
    D. Neither A nor B

    (D1.1)

122. A customer has requested a snowplow to be installed on his pickup truck. Technician A says that the plow kit should only be installed on vehicles approved for plowing. Technician B says that the headlight may need to be adjusted after installation has been completed. Who is correct?
    A. A only
    B. B only
    C. Both A and B
    D. Neither A nor B

    (D1.1)

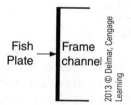

123. Technician A says the frame pictured shows a Fish Plate on the outside of the frame. Technician B says that the Fish Plate typically runs the full length of the chassis. Who is correct?
    A. A only
    B. B only
    C. Both A and B
    D. Neither A nor B

    (D1.3)

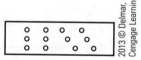

124. In the figure shown a series of holes need to be drilled. Technician A says that the holes should be drilled in line with each other to ease installation. Technician B says that the diagonal pattern will allow for larger diameter bolts to be used. Who is correct?
    A. A only
    B. B only
    C. Both A and B
    D. Neither A nor B

    (B1.3)

125. Technician A says that wood spacers are not required when using Shear Plates to mount a platform. Technician B says that Shear Plates and "U" bolts maybe combined to mount a body. Who is correct?
    A. A only
    B. B only
    C. Both A and B
    D. Neither A nor B

    (D1.2)

126. The customer requires a higher body deck height to allow their forklift to unload the platform. Technician A says that tubular steel can be stacked under the body long sill to raise body. Technician B says that the center of gravity should be reviewed to ensure the application meets the regulations. Who is correct?
    A. A only
    B. B only
    C. Both A and B
    D. Neither A nor B   (D1.2)

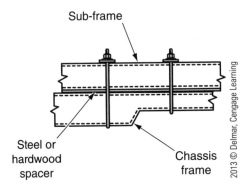

127. Technician A says that some factory frames have rivet heads in the flange and the wood spacer allows for proper crush on the "U" bolts. Technician B says that some trucks use both dual and single frame rails and using wood spacers can make up the gap. Who is correct?
    A. A only
    B. B only
    C. Both A and B
    D. Neither A nor B   (D1.2)

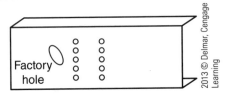

128. Several holes need to be drilled into a frame rail next to some existing factory holes. Which of these are correct?
    A. The factory hole should be plug welded and ground smooth
    B. The new hole should be placed 2 inches away to prevent cracking
    C. The new hole should be placed 1 ½ times the diameter from the edge of the smaller hole
    D. The new hole should be placed 1 ½ times the diameter from the center of the larger hole   (D1.2)

129. Technician A says that using fasteners with serrated edges on truck frames will help retain the torque of the bolt. Technician B says that that using a torque wrench will help ensure that the bolt was tightened to specs. Who is correct?
    A. A only
    B. B only
    C. Both A and B
    D. Neither A nor B   (D1.4)

TIRE AND LOADING INFORMATION

SEATING CAPACITY | TOTAL | FRONT | REAR

The combined weight of occupants and cargo should never exceed ___ Kg or Lbs

TIRE	SIZE	COLD TIRE PRESSURE	
FRONT			SEE OWNER'S MANUAL FOR ADDITIONAL INFORMATION
REAR			
SPARE			

130. Technician A says that the decal shown needs to be updated after the vehicle has been weighed. Technician B says the decal below must be changed if the seating has been altered. Who is correct?
  A. A only
  B. B only
  C. Both A and B
  D. Neither A nor B   (D2.9)

131. A wet line kit is being installed on the curbside of the vehicle. Technician A says a guard should be installed to protect the system from the exhaust. Technician B says that a decal should be added to alert the operator of the operation of the system. Who is correct?
  A. A only
  B. B only
  C. Both A and B
  D. Neither A nor B   (D2.6)

132. Technician A says that safety glasses should be worn when venting the refrigerant from the vehicle. Technician B says that the manufacturer's information should be reviewed before tapping into the system. Who is correct?
  A. A only
  B. B only
  C. Both A and B
  D. Neither A nor B   (D2.7)

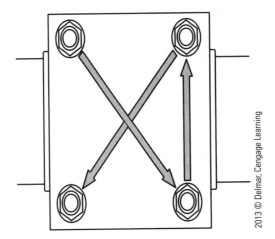

133. The pattern shown is used for installing "U" bolts on a leaf spring assembly. Technician A says that a step up torque procedure should be used in this application. Technician B says that the vehicle should be sitting at curb height when performing this procedure. Who is correct?
    A. A only
    B. B only
    C. Both A and B
    D. Neither A nor B

(B1.3)

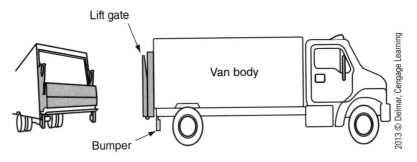

134. A van body with a liftgate has been installed on a class 7 chassis. Technician A says the liftgate is more than 24 inches away from the rear tires, so a compliant Interstate Commerce Commission Bumper will need to be installed. Technician B says that the liftgate will need to be measured as part of the overall length of the vehicle for legal compliance. Whose is correct?
    A. A only
    B. B only
    C. Both A and B
    D. Neither A nor B

(D2.5)

135. A customer complains of a slapping type sound when the truck is driven off road. Technician A says that an elongated hole in the frame holding a shear plate could be the cause. Technician B says the Shear Plate weld may have cracked and could be the cause for this sound. Who is correct?
    A. A only
    B. B only
    C. Both A and B
    D. Neither A nor B

(B2.2)

136. Technician A says that the Interstate Commerce Bumper (ICC) standard applies to trailer applications also. Technician B says that the ICC bumper prevents damage from occurring to the vehicle. Who is correct?
   A. A only
   B. B only
   C. Both A and B
   D. Neither A nor B

(B2.5)

137. Technician A says that when working near the exhaust system brackets shields should be used to protect the components installed. Technician B says that when installing the body the Technician should review the requirements for Fuel System integrity information from the manufacturer. Who is correct?
   A. A only
   B. B only
   C. Both A and B
   D. Neither A nor B

(B2.6)

138. When tapping into an air condition system the Technician should do all of these actions EXCEPT:
   A. Wear proper safety protection
   B. Replace gas only with the new EPA approved version
   C. Evacuate and recycle the old refrigerant where possible
   D. Measure and replace the oil taken from the system

(B2.7)

139. Technician A says that the operation decals supplied by a liftgate supplier help the customer operate the equipment. Technician B says that when possible the Technician should review the operation of the liftgate with the customer. Who is correct?
   A. A only
   B. B only
   C. Both A and B
   D. Neither A nor B

(D2.8)

7 Appendices

Answers to the Test Questions for the Sample Test Section 5

1. B	13. B	25. B	37. A
2. D	14. A	26. B	38. B
3. A	15. B	27. B	39. B
4. C	16. C	28. C	40. B
5. C	17. C	29. C	41. B
6. A	18. C	30. B	42. B
7. A	19. B	31. A	43. A
8. C	20. A	32. C	44. A
9. B	21. D	33. C	45. C
10. A	22. D	34. A	
11. C	23. B	35. A	
12. A	24. A	36. C	

Explanations to the Answers for the Sample Test Section 5

Question #1
Answer A is incorrect. This is the cab-to-axle (CA) dimension.
Answer B is correct. The wheelbase dimension is axle centerline to axle centerline on a single axle truck.
Answer C is incorrect. There is no such published dimension.
Answer D is incorrect. There is no such published dimension.

Question #2
Answer A is incorrect. This information is found in several documents and is not exclusive to the body builder.
Answer B is incorrect. There is no fixed dimension form the back of the cab.
Answer C is incorrect. The front of body does not have to be at the exact location of the published usable cab-to-axle.
Answer D is correct. This is the actual amount of linear distance from the closest point to the cab to the rear axle centerline that a body can begin.

Question #3
Answer A is correct. Only Technician A is correct. A Technician should review the Body Builders Drawing furnished by the chassis manufacturers and verify that they match the chassis being worked on.
Answer B is incorrect. The vehicle VIN tag does not contain dimensional data.
Answer C is incorrect. The Final Vehicle Manufacturers tag is installed upon the completion of the truck.
Answer D is incorrect. There is no reason to check this dimension.

Question #4
Answer A is incorrect. There is no absolute dimension of tandem spreads.
Answer B is incorrect. As with A, there is no absolute dimension of tandem spreads.
Answer C is correct. Tandem axles distances are part of the Bridge Weight Formula.
Answer D is incorrect. This dimension may be provided where a non-driven lift or helper axle are installed.

Question #5
Answer A is incorrect. Either the front or rear axle weight rating cannot be exceeded legally.
Answer B is incorrect. Either the front or rear axle weight rating cannot be exceeded legally.
Answer C is correct. All published weight rating must be within specifications for the vehicle to be compliant.
Answer D is incorrect. The correct payload if not placed correctly on the chassis could overload an axle.

Question #6
Answer A is correct. Body Builder's Manuals published by the chassis manufacturers will contain spring deflection information and ride height calculations. These calculations will include tire deflection as well.
Answer B is incorrect. The position of the helper spring pads is not relevant to this calculation.
Answer C is incorrect. There is no way to accurately load the vehicle with a simulated weight.
Answer D is incorrect. There is no "rule of thumb" for spring deflection. Various types of springs will deflect at different rates.

Question #7
Answer A is correct. DPFs are located in a specific location by the manufacturer that ensures that emission regulations are compliant. The distance from the engine discharge point to the DPF is a calculated dimension that allows for heat loss through the pipe leading to the DPF.
Answer B is incorrect. It is a violation of EPA emission standards to alter a tail pipe diffuser as this item mitigates the hot gases.
Answer C is incorrect. Only Technician A is correct.
Answer D is incorrect. Only Technician A is correct.

Question #8
Answer A is incorrect. While it is customary to vent batteries that are mounted in completely enclosed locations, batteries that are mounted outside of the truck or under the hood are not required to have distinct venting systems.
Answer B is incorrect. The high current circuit breakers do not have to be mounted in the battery box.
Answer C is correct. Federal Motor Vehicle Commercial Carrier Regulations require that the batteries are protected from above by a "substantial" cover. The definition of "substantial" is vague.
Answer D is incorrect. There is no regulation requiring a structural partition between the batteries.

Question #9
Answer A is incorrect. Technician B is correct. Crossmembers should be treated as heat-treated frame rails so welding or applying heat to them can alter its strength.
Answer B is correct. Only Technician B is correct. Welding to heat-treated frame can alter its hardness and truck crossmembers should be treated the same way as frame rails.
Answer C is incorrect. Only Technician B is correct.
Answer D is incorrect. Only Technician B is correct.

Question #10
Answer A is correct. When enough force is applied to a given point on a frame, the frame will take on a permanent "bend." The amount of force needed to bend the frame is the yield strength.
Answer B is incorrect. There is no shear strength measurement of a frame.
Answer C is incorrect. The RBM (Resistance Bending Moment) is an indication of the *overall* strength of the frame.
Answer D is incorrect. The volume of steel at a given point on the frame is the Section Modulus.

Question #11
Answer A is incorrect. Technician B is also correct.
Answer B is incorrect. Technician A is also correct.
Answer C is correct. Both Technicians are correct. Even a crack that extends into the radius of the frame can be repaired and a Fish Plate can be installed to correct the issue.
Answer D is incorrect. Both Technicians are correct.

Question #12
Answer A is correct. The ride height of a truck is one of many dimensions used to calculate driveline angles by the chassis manufacturer. A ride height that is too high or too low can cause driveline angles to become too severe and eventually cause failures in universal joints and other components.
Answer B is incorrect. The location of the valve is of no relevance in to this potential failure.
Answer C is incorrect. Mismatched parts will not cause this type of failure.
Answer D is incorrect. Answer A is correct.

Question #13
Answer A is incorrect. Dump valves can be used in all air suspension equipped vocations.
Answer B is correct. This is one of the key operational characteristics of an air ride suspension system. A dumped system will significantly improve the loading of the truck and add stability when a heavy lift truck is used.
Answer C is incorrect. The air suspension does not have to be dumped for extended storage of the vehicle.
Answer D is incorrect. The ride height control valve in effect controls this operation.

Question #14
Answer A is correct. Most OEMs require the use of ballast in snowplow applications.
Answer B is incorrect. While this may help to raise the front ride height, it is usually not recommended by the OEM.
Answer C is incorrect. The bracket was designed for the chassis and any alteration could cause severe damage to chassis frame.
Answer D is incorrect. A wheel alignment will not alter the ride height of the front end.

Question #15
Answer A is incorrect. It is not good practice to replace a set of springs only on one side without reviewing underlying causes.
Answer B is correct. Only Technician B is correct. A thorough review of the suspension is warranted.
Answer C is incorrect. Only Technician B is correct.
Answer D is incorrect. Only Technician B is correct.

Question #16
Answer A is incorrect. While low air pressure is a possibility, it is also unlikely. Low pressure would cause a reduction in the rate of fill of the airbag, but not affect the initial ride height.
Answer B is incorrect. A leaking air spring would cause the truck to lean. In this case the ride height is level, just low.
Answer C is correct. The height control valve is responsible to maintain the ride height and in as much as it is totally dependent on proper adjustment to function correctly, this is the first place to start in the diagnostic procedure.
Answer D is incorrect. The pressure regulator valve is part of the air compressor system, and not the air ride system.

Question #17
Answer A in incorrect. Center bearings are dictated by length of wheelbase and maintaining a good driveline angle also.
Answer B is incorrect. Depending on the torque, shaft speed and other factors the driveline will vary greatly.
Answer C is correct. In designing the truck, the original chassis manufacturer selected a driveshaft maximum based on shaft speeds, shaft thickness, and other studies.
Answer D is incorrect. While the length of the driveline may be part of the formula to determine the type of driveshaft used it is not the only factor.

Question #18
Answer A is incorrect. Answer C is correct.
Answer B is incorrect. Answer C is correct.
Answer C is correct. Thickness is selected by the original chassis manufacturer based on speeds, lengths, and critical speed calculations of the driveline in the truck.
Answer D is incorrect. Answer C is correct.

Question #19
Answer A is incorrect. A harmonic noise at highway speeds would be very difficult to describe. An out of phase driveline would produce a vibration, not a noise.
Answer B is correct. Because the driveshaft has to make two operating cycles during each rotation, and because the phasing is incorrect, a very pronounced vibration will be experienced at all speeds and get worse with high speed.
Answer C is incorrect. A resonance is a sound, not a vibration. Phasing problems do not produce sounds.
Answer D is incorrect. Answer B is correct.

Question #20
Answer A is correct. Operating angles of traditional universal joints are designed to be within 1 to 3 degrees.
Answer B is incorrect. Answer A is correct.
Answer C is incorrect. Answer A is correct.
Answer D is incorrect. While the angles of operation are somewhat codependent of each other, one joint can be extreme and fail, while the others are all acceptable.

Question #21
Answer A is incorrect. A center pin would not cause this issue.
Answer B is incorrect. An over loaded front axle would not cause this issue.
Answer C is incorrect. An under inflated tire will not cause this issue.
Answer D is correct. A worn torque arm bushing could cause the axles to shift and change the pinion angle.

Question #22
Answer A is incorrect. An alignment maybe needed but not till the chassis is at normal ride height
Answer B is incorrect. The springs may need to be replaced due to the added weight but it cannot be determined till the vehicle has been completed.
Answer C is incorrect. Neither Technician is correct.
Answer D is correct. Neither Technician is correct.

Question #23
Answer A is incorrect. There is no reason to check the grounds prior to beginning work on the chassis. Upon completion of the truck, a Technician will do an electrical inspection to check the function of all components, and at that time they will inspect ground circuits only if there is a problem, or if they were disturbed during the truck assembly.
Answer B is correct. It is very important to isolate the electrical circuits on the truck as much as possible. The best way to do this is to disconnect the battery ground cable before starting any work. This is not just to prevent damage to electrical components. It also serves to provide a margin of safety by preventing accidental shocks and potential fires.
Answer C is incorrect. There is no purpose to this action.
Answer D is incorrect. While this may help protect the main fuse box itself, it will not protect the rest of the truck circuits.

Question #24
Answer A is correct. A Technician should begin work with an inspection of the vehicle. Be particularly aware of major modifications that may have been done by someone else, such as wheelbase modifications or the addition of auxiliary axles.
Answer B is incorrect. Never assume that everything is good. The extra time it takes to inspect the chassis can be the difference between a good final vehicle and one that is troublesome.
Answer C is incorrect. The Final Vehicle Certification Label is not installed until the completion of the vehicle and the final compliance inspection.
Answer D is incorrect. This operation will not be completely done by the chassis preparation department. While the preparation department may install a PTO and therefore place the operators manual in the cab, they will not have installed all other components of the truck that require an operators manual.

Question #25
Answer A is incorrect. The crush block has nothing to do with the rear under-ride protection system.
Answer B is correct. Chassis manufacturers require crush blocks to be installed and state this requirement in the Body Builder's Guide. The crush block acts to help prevent deformation of the flange on a C channel frame when the "U" bolts are torqued initially and again in the future.
Answer C is incorrect. While the crush blocks serve to protect the frame flanges, they do not strengthen them. Their strength in part of the original design is based on size and material strength.
Answer D is incorrect. "U" bolt stretch does happen to an extent when they are tightened and torqued. However, this will not diminish with the installation of crush blocks.

Question #26
Answer A is incorrect. The point of maximum spring deflection indicates that the truck is under load. The upper rebound limit can occur with or without load.
Answer B is correct. This occurrence is the point of maximum travel including the travel of the frame and the rear axle in opposite directions. At this point, the possibility of a rear axle component breaking is at or near the limit. Manufacturers specify the minimum clearance over the tires that they feel is permissible. This clearance must be adhered to and may require the use of wheel boxes in the design of the body being mounted.
Answer C is incorrect. While one of the rear bumpers may be touching the frame or axle housing during the upper rebound limit travel, the other may be a great distance from the axle housing or frame.
Answer D is incorrect. Trucks with air suspension also have upper rebound limits, but this is not exclusive to trucks with air bags.

Question #27
Answer A is incorrect. Cutting the insulator at 90 degrees will create a stress riser on the frame. This should be avoided.
Answer B is correct. The insulator should be cut at a 45-degree or greater angle to spread the load over a greater distance through the frame. Some manufacturers provide a formula to calculate this angle based on the thickness of the insulator itself.
Answer C is incorrect. A Technician should never taper a body crossmember for any reason.
Answer D is incorrect. Neither the insulator nor any other items should be bolted to a frame flange.

Question #28
Answer A is incorrect. The diameter of the bolt is only one part of the formula used to determine bolt strength.
Answer B is incorrect. The style and type of threads are also a part of the equation, but not the sole determining factor.
Answer C is correct. This is the single most determining factor in the equation to rate bolts, nuts and washers. The quality of the material used in a grade eight bolt, for example, is very high tensile strength steel that will withstand a high degree of tension and shear.
Answer D is incorrect. The bolt flange, while an important part of the bolt, is not a determining factor in bolt grade.

Question #29
Answer A is incorrect. Technician B is also correct.
Answer B is incorrect. Technician A is also correct.
Answer C is correct. Both Technicians are correct. A rivet can be replaced by a comparable size grade of fastener. Always refer to OEM guidelines for information. A fastener can be replaced by a comparable size grade of rivet. Always refer to OEM guidelines for information.
Answer D is incorrect. Both Technicians are correct.

Question #30
Answer A is incorrect. The front of a sub-frame is prepared at a 90-degree angle.
Answer B is correct. The sub-frame should support the entire load placed on or in the body of the truck.
Answer C is incorrect. Never weld to the frame of a truck.
Answer D is incorrect. "U" bolts should be placed 18–24 inches back from front of the body.

Question #31

Answer A is correct. A full sub-frame should be used whenever possible to support two body structures. This will help spread the load over a greater dimension and reduce stress risers considerably.
Answer B is incorrect. Constructing two sub-frames, one for each body structure, will create a stress riser in between the two members.
Answer C is incorrect. While this would support the dump body sufficiently, the toolbox would rest on the truck frame only. The front of the dump body would be well centered on the truck frame, creating a stress point in a location well behind the back of the cab. This may NOT be the strongest point of the frame.
Answer D is incorrect. This would provide a concentrated load at the hoist mounting area and the rear hinge area of the dump body would not be in the same load plane as the other components.

Question #32

Answer A is incorrect. Mounting the first "U" bolt in this area will stiffen the frame not allowing it to flex. This will decrease ride quality significantly, and create a possible point of frame failure.
Answer B is incorrect. Mounting the first "U" bolt in this area will stiffen the frame not allowing it to flex. This will decrease ride quality significantly, and create a possible point of frame failure.
Answer C is correct. The location of the first "U" bolt should be in front of the first crossmember where possible which is typically 18 inches back from front of body.
Answer D is incorrect. This location would be too far rearward and create an unsatisfactory mount.

Question #33

Answer A is incorrect. Technician B is also correct.
Answer B is incorrect. Technician A is also correct.
Answer C is correct. Both Technicians are correct. The "U" bolts could be replaced. A shear plate can be used in most situations as well.
Answer D is incorrect. Both Technicians are correct.

Question #34

Answer A is correct. The headboard, in this case, is a removable object, meaning that the truck could be operated without it so the light needs to be mounted to a permanent structure.
Answer B is incorrect. Amber is the correct lens color for forward facing lights.
Answer C is incorrect. Technician A is correct.
Answer D is incorrect. Technician A is correct.

Question #35

Answer A is correct. As per the Federal Lighting Guide a tractor must have white conspicuity tape applied at the highest and widest part of the rear of the truck.
Answer B is incorrect. White is specified by FMVSS.
Answer C is incorrect. The truck is too large to require a center high stop light per FMVSS.
Answer D is incorrect. A back up light is not mandatory per FMVSS.

Question #36

Answer A is incorrect. Technician B is also correct.
Answer B is incorrect. Technician A is also correct.
Answer C is correct. Both Technicians are correct. While there may be a request from the customer for a specific fuel fill location, the actual location will depend on the original equipment manufacturers tank location, and the body arrangement. The fuel fill rate should be part of reviewing work upon completion of work.
Answer D is incorrect. Both Technicians are correct.

Question #37

Answer A is incorrect. Technician B is also correct.
Answer B is incorrect. Technician A is also correct.
Answer C is correct. Both Technicians are correct. The angle of the fuel cup could cause fill rate issues. Newer fuel kits are designed to slow down the fill rate to reduce vapor emissions. Technician should review with customer.
Answer D is incorrect. Both Technicians are correct.

Question #38
Answer A is incorrect. The label should be in close proximity to the controls for device that it describes.
Answer B is correct. Placing the label near the lift gate operation control will insure that the operator has a clear vision and convenient location.
Answer C is incorrect. The Technician needs to apply the operation and warning decals.
Answer D is incorrect. The Technician needs to apply the operation and warning decals.

Question #39
Answer A is incorrect. The IVD is used to instruct and inform the final vehicle manufacturer about compliance of the incomplete vehicle to Federal Motor Vehicle Safety Regulations. This document does not go along with the truck upon completion.
Answer B is correct. A requirement of the National Highway Traffic and Safety Administration is that the IVD, bearing the VIN of the truck is kept on file at the final vehicle manufacturers location.
Answer C is incorrect. A part of the IVD can be discarded, but the portion showing the vehicle VIN, GVW, and GAWRs must be retained.
Answer D is incorrect. The IVD does not go to the owner of the truck. The placement of the Final Vehicle Manufacturers label on the truck is the owner's proof that the truck was completed to comply with all FMVSS requirements.

Question #40
Answer A is incorrect. In as much as there are no hoods on certain styles of truck, this is not a suitable location. Also, federal regulations do not allow for this mounting location.
Answer B is correct. The location of the doorpost/jam area complies with federal regulations that require the label to be located in a conspicuous location.
Answer C is incorrect. In as much as the door is a removable part of the truck, this is not an appropriate location for the label.
Answer D is incorrect. The label needs to be placed per federal regulations; the sun visor does not meet that requirement.

Question #41
Answer A is incorrect. No surge will be caused by an improper vent system. It is possible, however, that a closed off vent system will cause the vehicle to stall.
Answer B is correct. The vent system must function correctly to allow air to escape while the tank is being refueled. If air cannot escape, the flow of fuel into the tank will be impeded, and the fill rate will be slow.
Answer C is incorrect. Fuel vapors should be captured buy the OEM collection systems, the vent hose would not cause this issue.
Answer D is incorrect. An improperly installed vent line would not cause this issue.

Question #42
Answer A is incorrect. No conspicuity tape requirements exist for straight trucks.
Answer B is correct. Federal Lighting Requirements state that on straight trucks over 30 feet in overall length, a side marker light must be installed.
Answer C is incorrect. There is no provision for a side turn signal in the body of a 30-foot truck.
Answer D is incorrect. There is a requirement as outlined by the Federal Lighting Regulations.

Question #43
Answer A is correct. Only Technician A is correct. The LED does not draw much current, so the flasher may need to be replaced. Review with the OEM on the proper method to achieve this.
Answer B is incorrect. Due to the lower amperage to operate the lights it's most likely that the flasher needs to be replaced.
Answer C is incorrect. Only Technician A is correct.
Answer D is incorrect. Only Technician A is correct.

Question #44
Answer A is correct. Only Technician A is correct. In most cases, R134A and R404 are the two acceptable, and legal, refrigerants.
Answer B is incorrect. R12 has been banned due to its ozone depleting properties.
Answer C is incorrect. Only Technician A is correct.
Answer D is incorrect. Only Technician A is correct.

Question #45
Answer A is incorrect. Technician B is also correct. 30 inches is the maximum height a rear bumper on straight truck.
Answer B is incorrect. Technician A is also correct. The ICC Regulation states the requirement for a rear bumper on a straight truck.
Answer C is correct. Both Technicians are correct. The ICC regulations are used to determine the correct measurements of the rear bumper on a straight truck. So it is always best to review when working on a bumper. The ICC regulation for a rear bumper height states that a bumper can be placed 30 inches or lower from the edge of the body.
Answer D is incorrect. Both Technicians are correct.

Answers to the Test Questions for the Sample Test Section 6

1. A	31. C	61. A	91. D
2. C	32. A	62. A	92. D
3. C	33. D	63. B	93. A
4. B	34. D	64. C	94. A
5. A	35. B	65. D	95. C
6. A	36. C	66. A	96. C
7. C	37. C	67. B	97. B
8. D	38. D	68. B	98. B
9. A	39. B	69. D	100. A
10. D	40. A	70. A	101. B
11. B	41. D	71. A	102. C
12. A	42. C	72. A	103. A
13. A	43. A	73. B	104. C
14. D	44. A	74. D	105. D
15. C	45. D	75. C	106. B
16. D	46. C	76. B	107. D
17. C	47. C	77. B	108. C
18. A	48. C	78. A	109. A
19. B	49. A	79. C	110. C
20. D	50. B	80. C	111. A
21. D	51. D	81. C	112. C
22. B	52. D	82. B	113. C
23. C	53. B	83. A	114. A
24. C	54. C	84. A	115. C
25. A	55. C	85. C	116. D
26. C	56. D	86. A	117. B
27. C	57. A	87. A	118. B
28. B	58. B	88. C	119. B
29. C	59. A	89. C	120. A
30. B	60. A	90. B	121. A

122. C	127. C	132. B	137. C
123. A	128. C	133. C	138. B
124. D	129. B	134. C	139. C
125. B	130. C	135. C	
126. C	131. C	136. D	

Explanations to the Answers for the Sample Test Section 6

Question #1
Answer A is correct. The cab-to-axle dimension is measured from the rear most portion of the cab to the centerline of the rear axle.
Answer B is incorrect. The cab-to-axle dimension is used to determine body length capacity of a particular truck.
Answer C is incorrect. This is a body builder manual publication. It is not printed on a data tag.
Answer D is incorrect. This dimension is generally not required for body mounting.

Question #2
Answer A is incorrect. This dimension has no purpose.
Answer B is incorrect. This does not specify which axle, so it is not valid.
Answer C is correct. This is the cab-to-trunnion dimension. It is the measurement from the center of the rear axle system and the back of the cab.
Answer D is incorrect. This dimension is unnecessary for body mounting purposes.

Question #3
Answer A is incorrect. The dimension is from the front bumper, not the rear bumper, to the back of the cab.
Answer B is incorrect. This is not a valid dimension for body mounting purposes.
Answer C is correct. The bumper to back-of-cab dimension is generally published for the consumer.
Answer D is incorrect. The bumper to back-of-cab is not a part of the weight distribution analysis.

Question #4
Answer A is incorrect. While adding more payload capacity, the addition of an auxiliary axle will not double the payload capacity of a truck.
Answer B is correct. The effective wheelbase will now be from the front axle of the truck to the center of the two rear axles. This will effect weight distribution and will require a new calculation.
Answer C is incorrect. Auxiliary axles can be mounted either in front of or behind a drive axle.
Answer D is incorrect. As stated, they can be in front of or behind a drive axle.

Question #5
Answer A is correct. The after-frame dimension refers to the centerline of the rear axle to the end of the truck frame. This can be extended or shortened as required.
Answer B is incorrect. The after frame is part of all trucks, with or without addition equipment.
Answer C is incorrect. As stated, the after frame can be modified.
Answer D is incorrect. In many cases, this dimension is not even published.

Question #6
Answer A is correct. This is the maximum legal weight of the truck with everything and everyone in it. It includes a full tank of fuel as well.
Answer B is incorrect. This is commonly referred to as "tare weight."
Answer C is incorrect. This is the Gross Combination Weight Rating (GVWR) as published by the manufacturer.
Answer D is incorrect. This is not just for registration purposes. In some states a Technician can register the vehicle at a different weight than the GVWR.

Question #7
Answer A is incorrect. Class seven is 26,001 to 33,000 pounds Gross Vehicle Weight Rating (GVWR).
Answer B is incorrect. Class five is 16,501 pounds to 26,500 pounds GVWR.
Answer C is correct. Any truck with a GVWR of 33,001 pounds and higher is a class eight unit. There is no class beyond eight.
Answer D is incorrect. A single axle tractor can be rated as a class seven.

Question #8
Answer A is incorrect. This is not necessarily in the owner's manual.
Answer B is correct. The Incomplete Vehicle Document (IVD) always contains the Gross Vehicle Weight Rating (GVWR).
Answer C is incorrect. There are multiple vehicle data plates.
Answer D is incorrect. The MSO is the Manufacturers Statement of Origin, and while it may contain the GVWR, it is not a document that the body builder has access to.

Question #9
Answer A is correct. FMVSS 121 uses the Gross Vehicle Weight Rating (GVWR) as an important part of compliance. Trucks are tested at GVWR for stopping distance compliance amongst others.
Answer B is incorrect. The addition of a snowplow does not change the GVWR of the truck. The addition of any equipment with the exception of auxiliary axles will not change the GVWR either.
Answer C is incorrect. Truck tractors DO have GVWR ratings. The load of the trailer on the fifth wheel cannot produce a truck whose GVWR has been exceeded.
Answer D is incorrect. All on-road vehicles, even trailers, have a GVWR rating.

Question #10
Answer A is incorrect. Tare weight refers to the empty weight of a complete vehicle. It is usually associated with trailers.
Answer B is incorrect. The Gross Vehicle Weight Rating (GVWR) is the published load capacity of a truck towing a trailer.
Answer C is incorrect. The GVWR does NOT have to equal the individual axle ratings.
Answer D is correct. None of the above answers are correct. All these answers are independent ratings and do not figure into a combined value if they are totaled together can mislead the end-user of their available capacity.

Question #11
Answer A is incorrect. The suitable tire and wheel combination rating is listed in several locations.
Answer B is correct. All trucks must have a tire and wheel tag applied in the doorpost area.
Answer C is incorrect. This tag is required by federal regulations.
Answer D is incorrect. All on road powered vehicles have a tire and wheel tag affixed to them.

Question #12
Answer A is correct. Along with the Vehicle Identification Number, the Gross Vehicle Weight Rating (GVWR) and the Gross Axle Weight Ratings must be printed on the Final Vehicle Certification Label.
Answer B is incorrect. The body tag does not require the GVWR of the truck.
Answer C is incorrect. This information is not required on the body bill of sale.
Answer D is incorrect. The addition of an auxiliary axle will require an additional Altered Vehicle Label.

Question #13
Answer A is correct. The Final Vehicle Manufacturer can change the Gross Vehicle Weight Rating (GVWR) of the truck, but rarely will do this without written consent from the Original Equipment Manufacturer.
Answer B is incorrect. The Final Vehicle Manufacturer does not have to authorize this change. They are not the Final Vehicle Manufacturer. This type of approval is rarely given.
Answer C is incorrect. The individual axle weight ratings can stay the same, and the GVWR increased by the addition of an auxiliary axle.
Answer D is incorrect. The GVWR can be changed by the addition of an auxiliary axle.

Question #14
Answer A is incorrect. Adding a lift gate does not require a change in Rear Gross Axle Weight Rating (RGAWR).
Answer B is incorrect. A forklift-loaded truck does not require a change in the RGAWR.
Answer C is incorrect. A tow truck does not require a change in the RGAWR.
Answer D is correct. None of the above will require a change in RGAWR.

Question #15
Answer A is incorrect. Technician B is also correct. The Upfitter is responsible to for the Final Stage Compliance with Federal Regulations listed in the Incomplete Vehicle Document.
Answer B is incorrect. Technician A is also correct. A vehicle certified as a "truck" will comply with all FMVSS requirements if completed in accordance to the IVD.
Answer C is correct. Both Technicians are correct.
Answer D is incorrect. Both Technicians are correct.

Question #16
Answer A is incorrect. There is no particular location on a chassis that is the measuring point of frame height.
Answer B is incorrect. As stated in A, there is no specific location for this measurement.
Answer C is incorrect. As stated above, this is a specific location and cannot be recognized.
Answer D is correct. This dimension is a variable. While most OEMs publish an "estimated" frame height at the end of the frame in the Body Builder's Manual, they also publish an end height calculation guide. Shortening or lengthening the after frame of a vehicle can change this height.

Question #17
Answer A is incorrect Technician B is also correct. Body insulators can affect body distance off frame.
Answer B is incorrect. Technician A is also correct. The correct body height will allow the tires to jounce properly without hitting the body.
Answer C is correct. Both Technicians are correct.
Answer D is incorrect. Both Technicians are correct.

Question #18
Answer A is correct. Most trucks have 34 inch frame widths, but they should always be measured before performing any work.
Answer B is incorrect. While 11 inches is a common height for many truck chassis, it is not a standard.
Answer C is incorrect. Technician A is correct.
Answer D is incorrect. Technician A is correct.

Question #19
Answer A is incorrect. This is not a usable specification for any purpose.
Answer B is correct. This is amount of frame that is behind the cab and any other OEM components like the vertical exhaust.
Answer C is incorrect. The back of cab may not be the only item protruding past the rear of the cab.
Answer D is incorrect. The OEM may publish this data but due to variables on the factory floor it's always best to measure once chassis is available.

Question #20
Answer A is incorrect. Technician B is also correct. The proper use of the cutting torch would be acceptable at the end of the frame rail.
Answer B is incorrect. Technician A is also correct. The frame rail may need to be shortened depending on the body.
Answer C is correct. Both Technicians are correct.
Answer D is incorrect. Both Technicians are correct.

Question #21
Answer A is incorrect. Technician A is incorrect. The entire chassis needs to be reviewed; the body may not be the widest point.
Answer B is incorrect. Technician B is incorrect; while the body length can contribute to the weight distribution, the placement of the axles IE the wheelbase would more likely affect the final distribution.
Answer C is incorrect. Both Technicians are incorrect.
Answer D is correct. Neither Technician A nor Technician B is correct.

Question #22
Answer A is incorrect. While 40 feet may be the maximum in some states on some roads, this is not a specific regulation.
Answer B is correct. Some local ordinances will over rule state regulations.
Answer C is incorrect. The overall length of a vehicle includes all parts on the vehicle, including front and rear bumpers if so equipped.
Answer D is incorrect. There is no formula for an overall length vs. wheelbase ratio.

Question #23
Answer A is incorrect. Technician B is also correct. This will help identify any issues before the Technician works on the vehicle.
Answer B is incorrect. Technician A is also correct. This review will ensure the Technician has a thorough understanding of the OEM expectations.
Answer C is correct. Both Technicians are correct.
Answer D is incorrect. Both Technicians are correct.

Question #24
Answer A is incorrect. Technician B is also correct. The after treatment device may affect the useable frame and cause the body to be moved rearward. If this occurs, a review of the Gross Vehicle Weight and vehicle length needs to be reviewed.
Answer B is incorrect. Technician A is also correct. Some equipment that is exhaust and/or noise emission related is illegal to relocate. This would include Diesel Particulate Filters.
Answer C is correct. Both Technicians are correct.
Answer D is incorrect. Both Technicians are correct.

Question #25
Answer A is correct. However, while legal, this is a modification that should be discussed with the chassis manufacturer first. Increased back pressure in the system can cause a malfunction to the emission control system and a fault code will be recorded in the engine electronic control module.
Answer B is incorrect. The OEM could offer a modification process need to review with them before moving any equipment.
Answer C is incorrect. The displacement of the engine, large or small, will still require the same cautions, but not affect the decision to make this change.
Answer D is incorrect. If the truck a Technician is working on is certified as an on-road truck, the percentage of time it spends in off-road use is meaningless. The engine emissions must still comply with the original certification.

Question #26
Answer A is incorrect. There is no requirement for the chassis manufacturer to authorize this. However, if the tank is part of another certification, for example side impact on a truck that is under 10,000 pounds GVWR, compliance must be maintained.
Answer B is incorrect. There is no reason to check the lateral center of gravity because of fuel tank relocation to the opposite side of the truck.
Answer C is correct. As long as all FMVSS and DOT requirements are met, the tank can be relocated.
Answer D is incorrect. There is no regard to the manufacturing process of the tank itself.

Question #27
Answer A is incorrect. Technician B is also correct. Extra air tanks are needed if the axle has a braking system.
Answer B is incorrect. Technician A is also correct. Air brake tanks can and often are relocated.
Answer C is correct. Both Technicians are correct.
Answer D is incorrect. Both Technicians are correct.

Question #28
Answer A is incorrect. It is legal to relocate control modules. However, utmost care and handling must be instituted and harness' modifications are not recommended.
Answer B is correct. As stated above, this may be done, but with extreme care and only when there is no alternative.
Answer C is incorrect. MOST anti-lock brake control modules are not frame mounted. There are exceptions.
Answer D is incorrect. All commercial motor vehicles require anti-lock brake system.

Question #29
Answer A is incorrect. A torch should never be used to install a hole in a frame rail.
Answer B is incorrect. While common practice, a twist drill is not the preferred toll to use to pierce a frame.
Answer C is correct. There are numerous cutting tools (bits) available that will leave a perfect hole without flaws. It should be noted that these tools require less effort, last longer and run cooler than twist drills.
Answer D is incorrect. While a plasma cutter leaves a good edge, it is not designed for piercing a hole in a frame.

Question #30
Answer A is incorrect. A twist drill will wear out quickly when used to remove a rivet. This is due to the hardness of the rivet itself, and the crown.
Answer B is correct. A cutting broach is the fastest and most useful tool for removing rivets.
Answer C is incorrect. Never use a torch to perform any work on a frame.
Answer D is incorrect. A cut-off tool will remove the head, but have no effect on the stem.

Question #31
Answer A is incorrect. There is no value to draining the fuel tank.
Answer B is incorrect. Again, there is no value to grounding the chassis.
Answer C is correct. Isolating the batteries by removing the ground cable is a good habit to get into. The will help prevent damage to electrical system components and act a as protection to electronic control modules.
Answer D is incorrect. Disconnecting the alternator main harness will protect only the alternator.

Question #32
Answer A is correct. A Technician can weld to cold rolled low yield strength steel, but must follow the guidelines of the manufacturers Body Builder's Manual. Also, welds must not be on the flanges, the radius sections at the top and bottom of the web, and must stay well below the radii themselves.
Answer B is incorrect. While it is better to bolt to the frame, in the case the bracket was intended to be welded on by the manufacturer so modifying the design may not be appropriate.
Answer C is incorrect. Adding a Fish Plate would not correctly repair the issue.
Answer D is incorrect. Adding a second bracket may alter the design intent of the manufacturer and should not be done unless an engineering review is conducted.

Question #33
Answer A is incorrect. This is a requirement for all fuel lines, especially soft line, but it is not the only requirement.
Answer B is incorrect. The routing of the fuel lines is of course important, but it is not the only requirement.
Answer C is incorrect. Again, it is absolutely critical that new fuel lines be of equal diameter as the original. This is true of the fuel return line as well.
Answer D is correct. Each of the above considerations must be completed when replacing or lengthening of fuel lines.

Question #34

Answer A is incorrect. While soldering wires together is in itself not a bad practice, solder joints can add resistance, and are susceptible to corrosion. Soldered joints should be shrink wrapped with the proper size wrapping material.

Answer B is incorrect. OBD II testers cannot recalibrate a system; they can only read a fault code. Also, there is no reason to recalibrate the ABS system if the wiring is extended.

Answer C is incorrect. Both Technician A and Technician B are incorrect.

Answer D is correct. Both Technician A and Technician B are incorrect.

Question #35

Answer A is incorrect. A Technician can pierce holes in the frame web area, but not near the radius.

Answer B is correct. Never pierce a frame flange. The flange is a primary area of strength for the frame. Holes placed in the flange for any reason, will weaken the frame at the location of the hole.

Answer C is incorrect. Even flanges in the after-frame area are subject to loads from forklift loading or lift gate installation and operation. It is not advisable to pierce a hole in any location on the frame flange.

Answer D is incorrect. A Technician can pierce reasonable sized holes in crossmembers for mounting small brackets and similar items.

Question #36

Answer A is incorrect. Answer D is correct.
Answer B is incorrect. Answer D is correct.
Answer C is incorrect. Answer D is correct.

Answer D is correct. Where available, the first choice should always be to use hardware included with the equipment being mounted, or hardware recommended by the equipment manufacturer.

Question #37

Answer A is incorrect. Technician B is also correct.
Answer B is incorrect. Technician A is also correct.

Answer C is correct. RBM is the representation of the overall strength of the frame. So by applying heat to the frame by welding can reduce its strength, while adding a Fish Plate can enhance the strength of the frame.

Answer D is incorrect. Both Technicians are correct.

Question #38

Answer A is incorrect. A Technician will not be able to keep all crossmembers in their original location, and will in fact, have to eliminate some. Crossmember placement is important and should be laid out carefully.

Answer B is incorrect. While this is one way of shortening, it is not the only method. In some cases, cutting and splicing the frame may be a better option.

Answer C is incorrect. As stated, a Technician can also "slide" the rear axle forward.

Answer D is correct. Either method may be used, based on the actual truck being worked on.

Question #39

Answer A is incorrect. Technician B is also correct. Welds are to be spaced apart (skip welded) in order to provide the weld with some flexibility.

Answer B is incorrect. Technician A is also correct. The diagram represents an OEM method to add an "L" section of frame to a chassis to increase strength in a high stress area on a truck frame.

Answer C is correct. Both Technicians are correct. The picture does represent the proper method to install an "L" reinforcing plate and the welds are staggered to prevent stress concentrations.

Answer D is incorrect. Both Technicians are correct.

Question #40
Answer A is correct. Frame tapers occur at the rear of the truck chassis, usually at or near the front spring hanger of the rear suspension. The best way to modify this style of frame for lengthening is to install a frame extension within the wheelbase of the truck, and reinforce the frame significantly.
Answer B is incorrect. Sliding the rear axle and its suspension rearward would actually place the suspension components on the weakest area of the frame. In addition, if the taper is small enough, there may be no physical space to mount the rear suspension rear hanger.
Answer C is incorrect. While not a great practice, it is possible to lengthen the wheelbase of a tapered frame truck.
Answer D is incorrect. The actual yield strength of the frame is not of real relevance to the action, other than knowing what welding can be done and how.

Question #41
Answer A is incorrect. The thickness of frame does not change the yield strength. Reinforcement sections do change the Section Modulus.
Answer B is incorrect. The published yield strength is equal throughout the frame.
Answer C is incorrect. Both Technicians are incorrect.
Answer D is correct. Both Technicians are incorrect.

Question #42
Answer A is incorrect. Toe is adjusted at the tie rod.
Answer B is incorrect. Camber is generally formed into the axle during assembly. Bending can only change it.
Answer C is correct. Caster can be changed by changing a wedge shaped shim mounted between the axle and the leaf spring on a straight axle truck.
Answer is incorrect. Wheel cut angle can be set at the wheel stop blocks or pads.

Question #43
Answer A is correct. This is a calculated value that is a real indicator of the overall strength of a frame rail.
Answer B is incorrect. Fish Plating of a frame is not needed unless the frame RBM and Section Modulus are insufficient.
Answer C is incorrect. Technician A is correct.
Answer D is incorrect. Technician A is correct.

Question #44
Answer A is correct. An aluminum frame offers weight savings to a customer.
Answer B is incorrect. An aluminum frame is no more or less flexible than a steel frame.
Answer C is incorrect. Technician A is correct. An aluminum frame offers weight savings to a customer but does not add any increased flexibility.
Answer D is incorrect. Technician A is correct. An aluminum frame offers weight savings to a customer but does not add any increased flexibility.

Question #45
Answer A is incorrect. While cracks in the frame web or flanges are important, they are not the only two points of inspection.
Answer B is incorrect. Crossmember cracks are unusual, and may be a result of an accident. However, they are not the only areas that should be inspected.
Answer C is incorrect. Cracks between holes in a frame are important to look for, but not exclusive to a good frame inspection.
Answer D is correct. A good frame inspection includes all of the above points and some additional inspection for straightness, sway, alignment, etc.

Question #46
Answer A is incorrect. This type of crack is repairable. Look to see if the holes were placed in the frame too close to one another as this can cause this type of failure.
Answer B is incorrect. This is not a proper method to repair this type of damage. This may have been a good idea before the truck was completed, but this is not a suitable method of repair at this point.
Answer C is correct. Proper welding procedures would include grinding the crack into a "V" notch shape that extends beyond the two holes and welding the area insuring that good penetration is achieved.
Answer D is incorrect. There is no need to do this if the welded repair is done correctly.

Question #47
Answer A is incorrect. Temperatures of the Diesel Particulate Filter (DPF) housing will exceed 500 degrees Fahrenheit. A 1-inch clearance may not be suitable.
Answer B is incorrect. While 3 inches may sound reasonable, for certain materials it may in fact be too close.
Answer C is correct. The chassis manufacturers Body Builder's Manual will show recommended clearances for all types of materials. A Technician should use this as a guide in making clearance decisions.
Answer D is incorrect. This is an unacceptable practice for high quality work. While some decisions can be based on cost effectiveness, this is a safety concern and that must come first in the decision-making.

Question #48
Answer A is incorrect. Technician B is also correct. The welding rod needs to exceed the rating of the frame rail.
Answer B is incorrect. Technician A is also correct. The only way to guarantee adequate weld penetration is to grind a "V" notch the length of the crack or longer, and then weld it completely.
Answer C is correct. Both Technicians are correct. The welding rod needs to be stronger than the frame material and then to ensure an adequate weld penetration, grind a "V" notch the length of the material to ensure a good application of weld.
Answer D is incorrect. Both Technicians are correct. The welding rod needs to be stronger than the frame material and then to ensure an adequate weld penetration, grind a "V" notch the length of the material to ensure a good application of weld.

Question #49
Answer A is correct. The rule of carpentry, "measure twice, cut one" applies to truck frames in this case. The best way to establish a pattern is to lay it out on something other than the frame itself first. This could be a light gauge sheet metal panel, or even a piece of cardboard. Once a Technician is certain that the pattern is correct, apply the template to the truck frame.
Answer B is incorrect. The level plane of the frame is not important in this case.
Answer C is incorrect. This may not be possible due to interference with other existing hardware, bolts, nuts, brackets etc.
Answer D is incorrect. Heating the frame will damage the frame and not make drilling easier.

Question #50
Answer A is incorrect. ¾ inches is the maximum size of hole that can be drilled in the frame.
Answer B is correct. To reduce the stress on the frame rail it's best to drill as close to the center of the frame as possible.
Answer C is incorrect. Technician A is correct.
Answer D is incorrect. Technician A is correct.

Question #51
Answer A is incorrect. 44,000 psi will be the easiest to pierce.
Answer B is incorrect. 56,000 psi is cold rolled, but still easy to pierce with the correct tools.
Answer C is incorrect. Most steel that is cold rolled is not heat treated and therefore not difficult to pierce.
Answer D is correct. 111,000 psi is heat-treated steel used mostly in class eight trucks. It is very hard steel and will require the use of special cutting tools to pierce a hole.

Question #52
Answer A is incorrect. This is the principle reason that we disconnect batteries prior to working on a truck, but it is not the sole reason.
Answer B is incorrect. This is another reason to disconnect the batteries, but it is not the sole reason.
Answer C is incorrect. Technicians will protect themselves by maintaining a clean, safe work area, but disconnecting the batteries alone will not protect them.
Answer D is correct. Each of the above actions will be realized by disconnecting the batteries of the truck being worked on.

Question #53
Answer A is incorrect. In fact, the chassis manufacturer does not want a Technician to disconnect the harness. These harness connectors are often weather guarded, and sealed. Some are equipped with very complex harness connectors and should not be disturbed.
Answer B is correct. It is adequate to remove the battery ground cable when welding to a truck.
Answer C is incorrect. Only Technician B is correct.
Answer D is incorrect. Only Technician B is correct.

Question #54
Answer A is incorrect. Technician B is also correct. The battery should be disconnected for safety reasons.
Answer B is incorrect. Technician A is also correct. Jack Stands should always be used for safety reasons.
Answer C is correct. Both Technicians are correct. Both Technicians are using good safety practices while working under the vehicle to perform a weld.
Answer D is incorrect. Both Technicians are correct. Both Technicians are using good safety practices while working under the vehicle to perform a weld.

Question #55
Answer A is incorrect. The starter motor is the least likely component to be damaged due to its construction and architecture.
Answer B is correct. The ECM is the most vulnerable as it is basically designed to work with milliamps, not high voltage.
Answer C is incorrect. Circuit breakers or fuses protect the instrument panel lights.
Answer D is incorrect. The batteries are rarely if ever damaged in this manner.

Question #56
Answer A is incorrect. It is not likely that a truck will leave a factory with a damaged frame.
Answer B is incorrect. Only because it is only one of the possibilities listed.
Answer C is incorrect. As above, only because it is only one of the possibilities listed.
Answer D is correct. Damage to a new truck frame prior to its arrival at the body manufacturers site is most likely a result of shipping damage. Both Answer B and C are the most likely reasons for frame damage. It is important that each truck chassis being received be inspected immediately upon arrival at the place of business.

Question #57
Answer A is correct. A Technician can and should check a frame for squareness prior to starting work. A tape measure is a readily available tool for this purpose, as long as a Technician is measuring exactly the same points on each side of the truck. Squareness measurements are diagonal from left to right side and then right to left side.
Answer B is incorrect. A line of site measurement is not accurate.
Answer C is incorrect. A string can be a good tool to measure the linear straightness of the frame rail, but it is not accurate for measuring squareness.
Answer D is incorrect. Unless the straightedge is as long as the truck, a Technician can only examine short distances, and this may not provide an accurate result.

Question #58
Answer A is incorrect. This is not a likely cause of this failure.
Answer B is correct. Improper torque and habitual looseness of a component that is "working" will elongate holes in a frame.
Answer C is incorrect. Yield strength has little to do with distortion of holes in a frame.
Answer D is incorrect. Corrosion will severely damage a frame over time, but cannot be specifically blamed for an elongated hole.

Question #59
Answer A is correct. If a crack in a crossmember is discovered it must be replaced.
Answer B is incorrect. Welding a crossmember is not the appropriate repair.
Answer C is incorrect. Replace the member with only a similar part.
Answer D is incorrect. Welding a crossmember is not the appropriate repair.

Question #60
Answer A is correct. A wheelbase discrepancy on each side of the truck requires further inspection. Some manufacturers provide a maximum deviation specification for this.
Answer B is incorrect. A frame web crack will most likely be caused by stress or possibly poor equipment mounting decisions. It is not a sign a bent frame.
Answer C is incorrect. This is a front-end alignment problem, not a frame problem.
Answer D is incorrect. If a leaf spring center bolt breaks repeatedly, the problem is most likely loose "U" bolts. Check the torque spec used to tighten them.

Question #61
Answer A is correct. A quick and simple method to check straightness is a string attached as far forward and as far rearward as possible to the side of a frame. Any bowing of the frame will be evident by a gap between the string and the web. If possible, check each side of the truck. This is a preliminary inspection, and if a gap is found, further more sophisticated methods should be used.
Answer B is incorrect. A straight edge will show distortion, but, only in a short span area. A Technician could miss the problem with this inspection.
Answer C is incorrect. This measurement would prove nothing.
Answer D is incorrect. Using a visual inspection to solely make a determination of repair is not considered a sound practice. The true extent of damage or twists can easily be missed.

Question #62
Answer A is correct. This is an indication that the original build of the truck was faulty. It is never a proper procedure to grind a relief into a frame rail flange for any reason.
Answer B is incorrect. The frame is in need of repair. It has not been repaired.
Answer C is incorrect. A chassis manufacturer would not provide a frame "notch" that is cut square. A straight cut will be more likely to crack than one that is curved. It is very unlikely that this was a "factory installed notch."
Answer D is incorrect. A frame can be repaired with this type of damage.

Question #63
Answer A is incorrect. Technician B is correct. The amount of damage will be the determining factor in the repair of a bent frame and it may include other components attached to the frame.
Answer B is correct. Heat-treated frames can be repaired if the damage is not that severe. Always measure damage and refer to OEM guidelines when making this determination.
Answer C is incorrect. Technician B is correct.
Answer D is incorrect. Technician B is correct.

Question #64
Answer A is incorrect. A "tag" axle is mounted behind the drive axle and will effectively lengthen the wheelbase.
Answer B is incorrect. When any auxiliary axle is installed, it will effectively change the wheelbase of the truck.
Answer C is correct. A "pusher" axle is mounted ahead of the drive axle, thus shortening the wheelbase of the truck.
Answer D is incorrect. The addition of any auxiliary axle to a truck is done with the purpose of changing the GVWR to increase payload.

Question #65
Answer A is incorrect. This is one reason for the insulator installation, but not the only reason.
Answer B is incorrect. Again, this is only one reason for the installation of an insulator.
Answer C is incorrect. While this does add to the flexibility of the chassis, it is not the sole reason for the use of an insulator.
Answer D is correct. The insulators provide all of the above-mentioned functions when properly installed in the truck.

Question #66
Answer A is correct. The crossmember allows some flex in a frame and aids significantly to the stability of the truck chassis.
Answer B is incorrect. A Technician should not use the crossmembers to mount additional equipment to a truck. If additional equipment is required, purpose built mounting brackets should be fabricated that attach to the frame, not the crossmember.
Answer C is incorrect. Generally, battery boxes, fuel tanks and other chassis mounted equipment do not require dedicated crossmembers.
Answer D is incorrect. Static flexibility is not a consideration for most trucks.

Question #67
Answer A is incorrect. Attaching a crossmember to a flange would require that a hole be pierced in the flange. It is never advisable to pierce a frame flange.
Answer B is correct. The most suitable location to mount a crossmember to is the frame web area.
Answer C is incorrect. Not all crossmembers support these bearings.
Answer D is incorrect. B is the correct answer.

Question #68
Answer A is incorrect. Tubular crossmembers are used in some passenger cars and many race cars. They are uncommon in trucks with the exception of hanger-bearing crossmembers. These hanger-bearing members do not serve to provide stability to the truck chassis.
Answer B is correct. The alligator style crossmember is the most common in use today.
Answer C is incorrect. An I beam crossmember would be heavy and serve no purpose.
Answer D is incorrect. There are no crossmembers designated as "cross beam."

Question #69
Answer A is incorrect. Rear suspension hangers are generally cast or forged and cannot, and should not, be repaired.
Answer B is incorrect. The spring pivot bolt needs to be inspected, but not automatically replaced.
Answer C is incorrect. Neither Technician A nor Technician B is correct.
Answer D is correct. Neither Technician A nor Technician B is correct.

Question #70
Answer A is correct. Shock absorbers should not exhibit any leakage of the oil from within the shock absorber. This type of leakage indicates a failed shock absorber.
Answer B is incorrect. A leaking air bag would not cause this issue.
Answer C is incorrect. Technician A is correct. This type of leakage indicates a failed shock absorber.
Answer D is incorrect. Technician A is correct. This type of leakage indicates a failed shock absorber.

Question #71
Answer A is correct. When loaded, the helper spring will carry a high percentage of the load. A weak or broken helper spring will create an off-level stance only when the truck is loaded. If the truck leans when empty, the main spring is most likely at fault.
Answer B is incorrect. Just because a spring shows "sign of rust" does not mean it is weak or broken.
Answer C is incorrect. The rear suspension bushings do not deflect under any circumstance, and thus cannot cause this problem.
Answer D is incorrect. A shock absorber carries no load and will not cause this problem.

Question #72
Answer A is correct. As a spring "works," its horizontal length will grow and retract with all vertical movement. The spring shackles are designed to allow for this change in horizontal length.
Answer B is incorrect. The shackles do not "control" the horizontal plain of the spring they allow it.
Answer C is incorrect. There is no stabilization action designed into a spring shackle.
Answer D is incorrect. Most helper spring mounting systems do not include shackles.

Question #73
Answer A is incorrect. Bounce control in an upper rebound event only, would be a single acting shock absorber.
Answer B is correct. Double acting shock absorbers cushion suspension travel in both up and down cycles.
Answer C is incorrect. Shock absorbers have no function in the positioning of the rear axle.
Answer D is incorrect. Pinion angle does change during acceleration, but this is not related in any way to the shock absorber action or function.

Question #74
Answer A is incorrect. The king pin housing is not a sealed unit.
Answer B is incorrect. Grease coming out of the pivot bearing is not an indicator of a worn bushing.
Answer C is incorrect. Using higher viscosity grease will not prevent this event.
Answer D is correct. There is no failure. This is a sign of full and complete penetration.

Question #75
Answer A is incorrect. Technician B is also correct. The torque arm does retain alignment, but this is not its only purpose.
Answer B is incorrect. Technician A is also correct. The torque arm does control torque, but this is not its singular function.
Answer C is correct. The arm controls torque and maintains alignment of the rear axle assembly.
Answer D is incorrect. Both Technicians are correct.

Question #76
Answer A is incorrect. Camber is controlled by the original design of the axle and when it requires correction, the axle must be bent and cannot be otherwise adjusted.
Answer B is correct. Camber can cause tire wear.
Answer C is incorrect. Technician B is correct.
Answer D is incorrect. Technician B is correct.

Question #77
Answer A is incorrect. The measurement of just the driveshaft tube itself is meaningless.
Answer B is correct. The total shaft length includes the yokes.
Answer C is incorrect. This is the driveline overall length.
Answer D is incorrect. This is not true. However, when a Technician is ordering a driveshaft for a modified wheelbase unit, they must discuss the point of measurements with the driveshaft shop to be certain that both are describing the same measurement points.

Question #78
Answer A is correct. The yokes ends MUST be in alignment with each other. This is correct phasing. If they are not in alignment, severe torsional vibrations will occur.
Answer B is incorrect. This is the description of an out of phase driveshaft.
Answer C is incorrect. This is not possible. Components are balanced individually. This is not phasing.
Answer D is incorrect. This is not phasing. However, a Technician should know that the crankshaft inclination angle and the rear pinion angle are important factors used to determine driveline angles, not phasing.

Question #79
Answer A is incorrect. Technician B is also correct. The working angles could cause this issue.
Answer B is incorrect. Technician A is also correct. The driveline could be out of balance.
Answer C is correct. Both Technicians are correct.
Answer D is incorrect. Both Technicians are correct.

Question #80
Answer A is incorrect. Technician B is also correct.
Answer B is incorrect. Technician A is also correct.
Answer C is correct. Both Technicians are correct. This failure will start as a vibration and end as a failed universal joint.
Answer D is incorrect. Both Technicians are correct.

Question #81
Answer A is incorrect. While these two angles play a part of the driveline angle calculation, they are rarely the same due to design criteria.
Answer B is incorrect. Crankshaft angles often exceed 5 degrees.
Answer C is correct. Because the transmission and the engine are directly coupled, the transmission main shaft MUST be at the same angle as the crankshaft.
Answer D is incorrect. The crankshaft inclination angle will not change during operation of the truck.

Question #82
Answer A is incorrect. A failed universal joint will generally fail at the bearing locations, not at the cross shaft.
Answer B is correct. The grease-fitting hole must be positioned in a location that will cause rotational forces on the shaft to compress the hole, not pull on it.
Answer C is incorrect. A shock load will generally cause driveshaft breakage, twist, or total destruction of the weakest point in the system.
Answer D is incorrect. Incorrect operating angles will cause vibrations and ultimately lead to total destruction of a joint.

Question #83
Answer A is correct. A lack of lubrication could be the cause.
Answer B is incorrect. Replacing only the needle bearings would not be appropriate.
Answer C is incorrect. Technician A is correct.
Answer D is incorrect. Technician A is correct.

Question #84
Answer A is correct. Shock loads generally cause twist.
Answer B is incorrect. Constant starts or high gear ratios will cause clutch failures, but not driveshaft twist.
Answer C is incorrect. Installing yokes out of phase will cause torsional vibrations, not driveshaft twist.
Answer D is incorrect. A broken universal joint will not result in driveshaft twist.

Question #85
Answer A is incorrect. A straight edge will not produce an accurate result.
Answer B is incorrect. This too will not produce an accurate result.
Answer C is correct. The dial indicator is the correct tool to use and will produce exact results.
Answer D is incorrect. While a person balancing a driveshaft will most likely check straightness, this is not the only time a shaft should be checked.

Question #86
Answer A is correct. Critical speed failures result very soon after driveline installation on modified wheelbase trucks. Combinations of gear ratios and driveshaft wall thickness will cause immediate failures when not considered in shaft assembly.
Answer B is incorrect. A short driveshaft, while creating its own problems, will generally NOT create a critical speed failure.
Answer C is incorrect. These problems will cause vibration issues and universal joint failures, but not destruction of the driveshaft itself.
Answer D is incorrect. Neither of these issues will cause a critical speed failure.

Question #87
Answer A is correct. A crush block is used to enhance a "U" bolt and should be located directly in line with the "U" bolt.
Answer B is incorrect. There is no specific installation interval for a crush block or a "U" bolt.
Answer C is incorrect. There is no specific application of a "U" bolt and crush block.
Answer D is incorrect. A "U" bolt and crush block can be used in the after-frame area as well.

Question #88
Answer A is incorrect. Oak is not the best material. Oak, and any wood will deteriorate in time and this can cause shrinkage. Shrinkage will cause the crush block to loosen and fall out of the truck.
Answer B is incorrect. Any type of wood will react the same way over time, and thus no wood is suitable for crush block construction.
Answer C is correct. A fabricated steel crush block is the best material for this job. Adding a retaining hoop that the "U" bolt can run through will insure that blocks are never lost.
Answer D is incorrect. Hard polypropylene will deform with torque applied to the "U" bolt, and is thus not a suitable material.

Question #89
Answer A is incorrect. A grade five bolt has three radial lines on the top.
Answer B is incorrect. Most metric bolts are marked with a number system.
Answer C is correct. This is the marking of a grade eight bolt.
Answer D is incorrect. There are no markings on grade two bolts.

Question #90
Answer A is incorrect. A template can be made from thin gauge sheet metal, but care must be used when used as a piercing fixture.
Answer B is correct. Heavy cardboard will not hold up for repeated use.
Answer C is incorrect. This material is good for this purpose, and is readily available in craft stores.
Answer D is incorrect. Steel plate that can be clamped to the frame and used as a guide is the best choice for repeated use.

Question #91
Answer A is incorrect. Sub-frames can be constructed of box section steel, and often are, but this is not the only type of material that can be used.
Answer B is incorrect. "C" Channel steel is acceptable for sub-frame construction, but is not the only material that should be used.
Answer C is incorrect. "I" beams do not offer the physical properties and an accommodating shape for a sub-frame construction.
Answer D is correct. A sub-frame can be constructed of either material shape.

Question #92
Answer A is incorrect. It is absolutely not a good idea to weld a sub-frame directly to a truck frame. This will create frame weakness, poor ride quality, and eventually can lead to frame failure.
Answer B is incorrect. Avoid welding to the truck frame whenever possible, even the frame web area.
Answer C is incorrect. Never pierce a hole in the top flange of a frame. This weakens the frame significantly.
Answer D is correct. Fabricating a suitable mounting bracket and bolting it to the frame is the best way to mount a sub-frame.

Question #93
Answer A is correct. A sub-frame should be used to isolate the Aluminum body from the steel truck frame typically a wood sub-frame is suggested by the OEM.
Answer B is incorrect. Technician A is correct. The sub-frame should be made of wood or a material to reduce the chance of accelerated corrosion between dissimilar metals.
Answer C is incorrect. Technician A is correct. A sub-frame should be used to isolate the Aluminum body from the steel frame.
Answer D is incorrect. Technician A is correct. A sub-frame should be used to isolate the Aluminum body from the steel frame.

Question #94
Answer A is correct. Welding process will require disconnecting of the batteries to protect electrical system.
Answer B is incorrect. There is no need to re-flash the Multiplex System.
Answer C is incorrect. Technician A is correct.
Answer D is incorrect. Technician A is correct.

Question #95
Answer A is incorrect. FMVSS 209 regulates seat belt assemblies.
Answer B is incorrect. FMVSS 121 regulates air brake systems.
Answer C is correct. FMVSS 108 regulates lighting systems.
Answer D is incorrect. FMVSS 111 regulates rear vision.

Question #96
Answer A is incorrect. Technician B is also correct. The manufacturer kit should not be modified.
Answer B is incorrect. The fuel filler and hoses need to be protected from heat sources.
Answer C is correct. Both Technicians are correct.
Answer D is incorrect. Both Technicians are correct.

Question #97
Answer A is incorrect. Federal requirements specifically prohibit the installation of a fuel filler inside of a vehicle component of any type. The venting system is not relevant to this requirement.
Answer B is correct. A lockable fuel cap could be installed on a fuel filler cap to prevent tampering with system.
Answer C is incorrect. Technician B is correct. A lockable fuel cap could be installed on a fuel filler cap to prevent tampering with system.
Answer D is incorrect. Technician B is correct. A lockable fuel cap could be installed on a fuel filler cap to prevent tampering with system.

Question #98
Answer A is correct. The Final Vehicle Certification label must document the VIN.
Answer B is incorrect. The Final Stage Decal does not require the body information.
Answer C is incorrect. Technician A is correct. A Final Stage Certification must include the VIN of the chassis; the serial number of the body is not a requirement on this label.
Answer D is incorrect. Technician A is correct. A Final Stage Certification must include the VIN of the chassis; the serial number of the body is not a requirement on this label.

Question #99
Answer A is correct. This label is required and the OEM installs it at the time of manufacturing.
Answer B is incorrect. Only one label is required, and it is generally located on the doorpost area.
Answer C is incorrect. This label is required to be placed on the vehicle by the OEM.
Answer D is incorrect. This label is installed per the Federal Guidelines.

Question #100
Answer A is incorrect. A used truck should already have a Final Stage label; otherwise the used vehicle could not be registered or licensed by the motor vehicle agencies. Installing a new body on a used truck does not require a new label.
Answer B is correct. A completed vehicle that was altered before the first retail sale requires an Altered Decal. The Altered Label indicates to the governing agencies that the vehicle has been altered from its original design and the entity installing the new body ensures that it meets all Federal Motor Vehicle Guidelines from the date of manufacture of the vehicle.
Answer C is incorrect. Technician B is correct. An Altered Label must be installed on the vehicle.
Answer D is incorrect. Technician B is correct. An Altered Label must be installed on the vehicle.

Question #101
Answer A is incorrect. Technician A is incorrect. While it may increase the tire clearance, cutting the long sill may affect the design of the body structure.
Answer B is correct. Technician B is correct. A sub-frame may be the best option to increase the clearance.
Answer C is incorrect. Technician B is correct.
Answer D is incorrect. Technician B is correct.

Question #102
Answer A is incorrect. Technician B is also correct. The main reason to add an axle is to increase payload.
Answer B is incorrect. Technician A is also correct. Typically the additional axle is called the Tag in this location.
Answer C is correct. Both Technicians are correct.
Answer D is incorrect. Both Technicians are correct.

Question #103
Answer A is correct. Technician A is correct. This maybe the most efficient method.
Answer B is incorrect. Technician B is incorrect. Since the new truck is being done for the dealer it has not been through its first retail sale. Also the Decal that is required is a Final Stage Decal.
Answer C is incorrect. Technician A is correct.
Answer D is incorrect. Technician A is correct.

Question #104
Answer A is incorrect. The decal must be located on street side hood, unless approval is received from the EPA.
Answer B is incorrect. The decal adhesive must last 10 years.
Answer C is correct. The decal is not designed for removal without authorization.
Answer D is incorrect. The label does include the VIN.

Question #105
Answer A is incorrect. Technician A is incorrect. The truck is less than 30 feet so no extra lights are required.
Answer B is incorrect. Technician B is incorrect. Per FMVSS 108 the lights should be as high as practical. So they are not required to be located in corners of body.
Answer C is incorrect. Both Technicians are incorrect.
Answer D is correct. Neither Technician A nor Technician B is correct.

Question #106
Answer A is incorrect. Technician A is incorrect. This truck has dually tires and the Gross Vehicle Weight requires reflectors at the corners.
Answer B is correct. Technician B is correct. The regulation states that the body can act as rear end protection at this height from ground.
Answer C is incorrect. Technician B is correct.
Answer D is incorrect. Technician B is correct.

Question #107
Answer A is incorrect. Single rear wheel axle indicates a vehicle rating of less than 10,000 Gross Vehicle Weight so it requires the CHMSL.
Answer B is incorrect. The chassis width is less than 80 inches so it requires the CHMSL.
Answer C is incorrect. A vehicle rating of less than 10,000 Gross Vehicle Weight requires a CHMSL.
Answer D is correct. The dual rear wheels indicate a vehicle capable of a GVW of over 10,000 pounds so a CHMSL is not required.

Question #108
Answer A is incorrect. Technician A is incorrect. Technician B is also correct. The lights are spaced to provide indication of the vehicle type at night.
Answer B is incorrect. Technician B is incorrect. Technician A is also correct. A class eight truck exceeds the 10,000 pound Gross Vehicle Weight so it needs to have a Tri Light at rear.
Answer C is correct. Both Technicians are correct.
Answer D is incorrect. Both Technicians are correct.

Question #109
Answer A is correct. Technician A is correct. The Federal Motor Vehicle Safety Standard only mandates one back up light.
Answer B is incorrect. Technician B is incorrect. There is not decibel standard from FMVSS as long as it is audible.
Answer C is incorrect. Technician A is correct.
Answer D is incorrect. Technician A is correct.

Question #110
Answer A is incorrect. Technician B is also correct. The Web of the frame is generally the best area to drill holes.
Answer B is incorrect. Technician A is also correct. The Web area is generally 2 inches from frame flanges.
Answer C is correct. Both Technicians are correct.
Answer D is incorrect. Both Technicians are correct.

Question #111
Answer A is correct. Technician A is correct. Adding a Leaf spring will aid in drivability.
Answer B is incorrect. Technician B is incorrect. Adding a Leaf does not affect GVW rating.
Answer C is incorrect. Technician A is correct.
Answer D is incorrect. Technician A is correct.

Question #112
Answer A is incorrect. Technician B is also correct.
Answer B is incorrect. Technician A is also correct.
Answer C is correct. Both Technicians are correct.
Answer D is incorrect. Both Technicians are correct.

Question #113
Answer A is incorrect. Technician B is also correct.
Answer B is incorrect. Technician A is also correct.
Answer C is correct. Both Technicians are correct.
Answer D is incorrect. Both Technicians are correct.

Question #114
Answer A is correct. Technician A is correct. It is always best to review with the manufacturer for proper repair procedures.
Answer B is incorrect. Technician B is incorrect. The shock absorber does not support the payload.
Answer C is incorrect. Technician A is correct.
Answer D is incorrect. Technician A is correct.

Question #115
Answer A is incorrect. Technician B is also correct. The slip joint can act as a seal for the transmission output shaft housing.
Answer B is incorrect. Technician A is also correct. The slip joint allows the shaft to lengthen and shorten to compensate for the suspension movement.
Answer C is correct. Both Technicians are correct.
Answer D is incorrect. Both Technicians are correct.

Question #116
Answer A is incorrect. Technician A is incorrect. Replacing a U-Joint in normal maintenance does not alter the driveline phasing.
Answer B is incorrect. Technician B is incorrect. Greasing should be done only when the end caps are lock into their housings. Pre-greasing may cause the cap to pop off under the pressure from the grease.
Answer C is incorrect. Both Technicians are incorrect.
Answer D is correct. Neither Technician A nor Technician B is correct.

Question #117
Answer A is incorrect. Technician A is incorrect. 3 degrees is within the tolerance of this assembly.
Answer B is correct. Technician B is correct. A driveline made too long may create torsional issues.
Answer C is incorrect. Technician B is correct.
Answer D is incorrect. Technician B is correct.

Question #118
Answer A is incorrect. Incorrect Phasing could cause failure.
Answer B is correct. 5 degrees is outside the tolerance of this assembly.
Answer C is incorrect. A balance driveline would not cause this issue.
Answer D is incorrect. Over greasing can create seal failure and then expose seal to contamination.

Question #119
Answer A is incorrect. Technician A is incorrect. These assemblies are sealed and cannot be lubricated.
Answer B is correct. Technician B is correct. The bearing assembly should be perpendicular to the frame rail.
Answer C is incorrect. Technician B is correct.
Answer D is incorrect. Technician B is correct.

Question #120
Answer A is correct. Technician A is correct. The rubber carrier for the bearing could cause this issue.
Answer B is incorrect. Technician B is incorrect. The rubber boot is designed to cover the slip joint.
Answer C is incorrect. Technician A is correct.
Answer D is incorrect. Technician A is correct.

Question #121
Answer A is correct. Technician A is correct. 3 inches is the accepted distance in this class of vehicle for the collision requirement.
Answer B is incorrect. Technician B is incorrect. The manufacturer of the vehicle or the body will make that determination.
Answer C is incorrect. Technician A is correct.
Answer D is incorrect. Technician A is correct.

Question #122
Answer A is incorrect. Technician B is also correct. Headlight aim may be affected.
Answer B is incorrect. Technician A is also correct. A proper installation requires a chassis rated for the rigors of the added weight and demands on the suspension.
Answer C is correct. Both Technicians are correct.
Answer D is incorrect. Both Technicians are correct.

Question #123
Answer A is correct. Technician A is correct. The figure does show a Fish Plate.
Answer B is incorrect. Technician B is incorrect. Fish Plates are typically added in areas were a high stress on the frame may occur.
Answer C is incorrect. Technician A is correct.
Answer D is incorrect. Technician A is correct.

Question #124
Answer A is incorrect. Technician A is incorrect. The pattern of the drilled holes has no bearing on the diameter of the holes that can be drilled.
Answer B is incorrect. Technician B is incorrect. The vertical alignment of the drilled holes may cause stress points in the frame rail. It's always best if possible to offset the pattern.
Answer C is incorrect Both Technicians are incorrect.
Answer D is correct. Neither Technician A nor Technician B is correct.

Question #125
Answer A is incorrect. Technician A is incorrect. Technician B is correct. Shear Plates have no bearing what Body Spacers or Insulators are required.
Answer B is correct. Technician B is correct. Shear Plates and "U" bolts may be combined to mount bodies.
Answer C is incorrect. Technician B is correct.
Answer D is incorrect. Technician B is correct.

Question #126
Answer A is incorrect. Technician B is also correct. A CG should be calculated to ensure the vehicle remains in the envelope.
Answer B is incorrect. Technician A is also correct. Steel Tubing can be used to raise a body.
Answer C is correct. Both Technicians are correct.
Answer D is incorrect. Both Technicians are correct.

Question #127
Answer A is incorrect. Technician B is also correct. Wood spacer can make up the height variances in frames.
Answer B is incorrect. Technician A is also correct. Wood spacers can be used to cover rivet heads to the body can sit flat on the frame rail.
Answer C is correct. Both Technicians are correct.
Answer D is incorrect. Both Technicians are correct.

Question #128
Answer A is incorrect. Welding should be avoided on frame rails.
Answer B is incorrect. 2 inches is incorrect depending on the application.
Answer C is correct. To reduce the chance for cracking is this rule of thumb.
Answer D is incorrect. This distance though greater than answer C, it may adversely affect the application.

Question #129
Answer A is incorrect. Technician A is incorrect. Technician B is correct. The use of a serrated bolt may dig into the frame rail causing long-term corrosion and other failures.
Answer B is correct. Technician B is correct. A torque wrench will help ensure a properly tightened bolt.
Answer C is incorrect. Technician B is correct.
Answer D is incorrect. Technician B is correct.

Question #130
Answer A is incorrect. Technician B is also correct. The seating chart must be updated.
Answer B is incorrect. Technician A is also correct. The available payload must be updated.
Answer C is correct. Both Technicians are correct.
Answer D is incorrect. Both Technicians are correct.

Question #131
Answer A is incorrect. Technician B is also correct. All supplied operation decals should be installed.
Answer B is incorrect. Technician A is also correct. Guards should be installed to protect the operator and the equipment from heat.
Answer C is correct. Both Technicians are correct.
Answer D is incorrect. Both Technicians are correct.

Question #132
Answer A is incorrect. Technician A is incorrect. Refrigerant should never be vented to the atmosphere.
Answer B is correct. Technician B is correct. It is always best to review manufacturer recommendations.
Answer C is incorrect. Technician B is correct.
Answer D is incorrect. Technician B is correct.

Question #133
Answer A is incorrect. Technician B is also correct. It's best to torque all suspension components at curb height, as it would side when off the rack.
Answer B is incorrect. Technician A is also correct. Using a step up torque procedure will ensure proper seating of the "U" bolts.
Answer C is correct. Both Technicians are correct.
Answer D is incorrect. Both Technicians are correct.

Question #134
Answer A is incorrect. Technician B is also correct. The liftgate is a permanent fixture so it needs to be included.
Answer B is incorrect. Technician A is also correct. The ICC Rule covers standard trucks.
Answer C is correct. Both Technicians are correct.
Answer B is incorrect. Both Technicians are correct.

Question #135
Answer A is incorrect. Technician B is also correct. A cracked weld in the Shear Plate could cause this issue.
Answer B is incorrect. Technician A is also correct. An elongated hole could cause this issue.
Answer C is correct. Both Technicians are correct.
Answer D is incorrect. Both Technicians are correct.

Question #136
Answer A is incorrect. Technician A is incorrect. Trailers have another standard for rear impact protection.
Answer B is incorrect. Technician B is incorrect. The ICC bumper is intended to prevent another vehicle from under riding the vehicle it is not intended to protect the vehicle.
Answer C is incorrect Both Technicians are incorrect.
Answer D is correct. Neither Technician A nor Technician B is correct.

Question #137
Answer A is incorrect. Technician B is also correct.
Answer B is incorrect. Technician A is also correct.
Answer C is correct. Both Technician are correct.
Answer D is incorrect. Both Technician are correct.

Question #138
Answer A is incorrect. The Technician should wear proper safety protection.
Answer B is correct. The Technician should replace with the gas required by the system as the newer gases operate at differing pressures and components. Replace gas only with the new EPA approved version.
Answer C is incorrect. The Technician should evacuate and recycle the old refrigerant where possible.
Answer D is incorrect. The Technician should measure and replace the oil taken from the system.

Question #139
Answer A is incorrect. Technician B is also correct. It is helpful to review with the customer the operation and maintenance of the liftgate.
Answer B is incorrect. Technician A is also correct. All the operational and safety decals should be installed.
Answer C is correct. Both Technician are correct.
Answer D is incorrect. Both Technician are correct.

Glossary

Actuator A device that delivers motion in response to an electrical signal.

AH (Ampere-Hours) An older method of determining a battery's capacity.

Alternator A device that converts mechanical energy from the engine to electrical energy used to charge the battery and power various vehicle accessories.

Ammeter A device (usually part of a DMM) that is used to measure current flow in units known as amps or milliamps.

Amp Clamp An inductive-style tool that can measure the current flow in a conductor by sensing the magnetic field around it at 9 degrees Fahrenheit.

Ampere A unit for measuring electrical current; also known as amp.

Analog Signal A voltage signal that varies within a given range from high to low, including all points in between.

Analog-to-Digital Converter (A/D converter) A device that converts analog voltage signals to a digital format, located in the section of a control module called the input signal conditioner.

Analog Volt/Ohmmeter (AVOM) A test meter used for checking voltage and resistance. These are older-style meters that use a needle to indicate the values being read; should not be used with electronic circuits.

Armature The rotating component of a (1) starter or another motor, (2) generator, (3) compressor clutch.

ATA Connector American Trucking Association data link connector; the standard connector used by most manufacturers for accessing data information from various electronic systems in trucks.

Blade Fuse A type of fuse having two flat male lugs for insertion into mating female sockets.

Blower Fan A fan that pushes air through a ventilation, heater, or air-conditioning system.

CCA (Cold Cranking Amps) A common method used to specify battery capacity.

CCM (Chassis Control Module) A computer used to control various aspects of driveline operation; usually does not include any engine controls.

Circuit A complete path for electrical current to flow.

Circuit Breaker A circuit protection device used to open a circuit when current in excess of its rated capacity flows through a circuit; designed to reset, either manually or automatically.

Data Bus Data backbone of the chassis electronic system using hardware and communications protocols consistent with CAN 2.0 and SAE J-1939 standards.

Data Link A dedicated wiring circuit in the system of a vehicle used to transfer information from one or more electronic systems to a diagnostic tool, or from one module to another.

Diode An electrical one-way check valve; allows current flow in one direction but not the other.

DMM (Digital Multi-meter) A tool used for measuring circuit values such as voltage, current flow, and resistance; has a digital readout, and is recommended for measuring sensitive electronic circuits.

ECM (Electronic Control Module) Acronym for the modules that control the electronic systems on a truck; also known as ECU (electronic control unit).

Electricity The flow of electrons through various circuits, usually controlled by manual switches and senders.

Electronically Erasable Programmable Memory (EEPROM) Computer memory that enables write-to-self, logging of failure codes and strategies, and customer/proprietary data programming.

Electronics The branch of electricity where electrical circuits are monitored and controlled by a computer, the purpose of which is to allow for more efficient operation of those systems.

Electrons Negatively charged particles orbiting every atomic nucleus.

EMI (Electro-Magnetic Interference) Low-level magnetic fields that interfere with electrical/electronically controlled circuits, causing erratic outcomes.

Fault Code A code stored in computer memory to be retrieved by a technician using a diagnostic tool.

Fuse A circuit protection device designed to open a circuit when amperage that exceeds its rating flows through a circuit.

Fuse Cartridge A type of fuse having a strip of low-melting-point metal enclosed in a glass tube.

Fusible Link A short piece of wire with a special insulation designed to melt and open during an overload; installed near the power source in a vehicle to protect one or more circuits, and is usually two to four wire gauge sizes smaller than the circuit it is designed to protect.

Grounded Circuit A condition that causes current to return to the battery before reaching its intended destination; because the resistance is usually much lower than normal, excess current flows and damage to wiring or other components usually results; also known as short circuit.

Halogen Light A lamp having a small quartz/glass bulb that contains a filament surrounded by halogen gas; is contained within a larger metal reflector and lens element.

Harness and Harness Connectors The routing of wires along with termination points to allow for vehicle electrical operation.

High-Resistance Circuits Circuits that have resistance in excess of what was intended. Causes a decrease in current flow along with dimmer lights and slower motors.

Inline Fuse A fuse usually mounted in a special holder inserted somewhere into a circuit, usually near a power source.

Insulator A material, such as rubber or glass, that offers high resistance to the flow of electricity.

Integrated Circuit A solid-state component containing diodes, transistors, resistors, capacitors, and other electronic components mounted on a single piece of material and capable of performing numerous functions.

IVR (Instrument Voltage Regulator) A device that regulates the voltage going to various dash gauges to a certain level to prevent inaccurate readings; usually used with bimetal type gauges.

Jump Start A term used to describe the procedure where a booster battery is used to help start a vehicle with a low or dead battery.

Jumper Wire A piece of test wire, usually with alligator clips on each end, meant to bypass sections of a circuit for testing and troubleshooting purposes.

Magnetic Switch The term usually used to describe a relay that switches power from the battery to a starter solenoid; is controlled by the start switch.

Maintenance-Free Battery A battery that does not require the addition of water during its normal service life.

Milliamp 1/1000th of an amp; 1000 milliamps = 1 amp.

Millivolt 1/1000th of a volt; 1000 millivolts = 1 volt.

Ohm A unit of electrical resistance.

Ohmmeter An instrument used to measure resistance in an electrical circuit, usually part of a DMM; power must be turned off on the electrical circuit before the ohmmeter can be connected.

Ohm's Law A basic law of electricity stating that in any electrical circuit, voltage, amperage, and resistance work together in a mathematical relationship; $E = I \times R$.

Open Circuit A circuit in which current has ceased to flow because of either accidental breakage (such as a broken wire) or intentional breakage (such as opening a switch).

Output Driver An electronic on/off switch that a computer uses to drive higher amperage outputs, such as injector solenoids.

Parallel Circuit An electrical circuit that provides two or more paths for the current to flow; each path has separate resistances (or loads) and operates independently from the other parallel path; in a parallel circuit, amperage can flow through more than one load path at a time.

Power A measure of work being done; in electrical systems, this is measured in watts, which is simply amps $\times$ volts.

Processor The brain of the processing cycle in a computer or module; performs data fetch-and-carry, data organization, logic, and arithmetic computation.

Programmable Read-Only Memory (PROM) An electronic memory component that contains program information specific to chassis application; used to qualify ROM data.

Random-Access Memory (RAM) The memory used during computer operation to store temporary information; the computer can write, read, and erase information from RAM in any order, which is why it is called random; RAM is electronically retained and therefore volatile.

Read-Only Memory (ROM) A type of memory used in computers to store information permanently.

Reference Voltage The voltage supplied to various sensors by the computer, which acts as a baseline voltage; modified by sensors to act as an input signal.

Relay An electrical switch that uses a small current to control a large one, such as a magnetic switch used in starter motor cranking circuits.

Reserve Capacity Rating The measurement of the ability of a battery to sustain a minimum vehicle electrical load in the event of a charging system failure.

Resistance The opposition to current flow in an electrical circuit; measured in units known as ohms.

Rotor (1) A part of the alternator that provides the magnetic fields necessary to generate a current flow; (2) the rotating member of an assembly.

Semiconductor A solid-state device that can function as either a conductor or an insulator depending on how its crystalline structure is arranged.

Sensing Voltage A reference voltage put out by the alternator that allows the regulator to sense and adjust the charging system output voltage.

Sensor An electrical unit used to monitor conditions in a specific circuit to report back to either a computer or a light, solenoid, etc.

Series Circuit A circuit that consists of one or more resistances connected to a voltage source so there is only one path for electrons to flow.

Series/Parallel Circuit A circuit designed so that both series and parallel combinations exist within the same circuit.

Short Circuit A condition, most often undesirable, in which two circuits, one circuit relative to ground or one circuit relative to another, connect; commonly caused by two wires rubbing together and exposing bare wires; almost always causes blown fuses and/or undesirable actions.

Signal Generators Electromagnetic devices used to count pulses produced by a reluctor or chopper wheel (such as teeth on a transmission output shaft gear), which are then translated by an ECM or gauge to display speed, rpm, etc.

Slip Rings and Brushes Components of an alternator that conduct current to the rotating rotor; most alternators have two slip rings mounted directly on the rotor shaft; they are insulated from the shaft and each other; a spring-loaded carbon brush is located on each slip ring to carry the current to and from the rotor windings.

Solenoid An electromagnet used to perform mechanical work, made with one or two coil windings wound around an iron tube; a good example is a starter solenoid, which shifts the starter drive pinion into mesh with the flywheel ring gear.

Starter (Neutral) Safety Switch A switch used to ensure that a starter is not engaged when the transmission is in gear.

Switch A device used to control current flow in a circuit; can be either manually operated or controlled by another source, such as a computer.

Transistor An electronic device that acts as a switching mechanism.

Volt A unit of electrical force or pressure.

Voltage Drop The amount of voltage lost in any particular circuit due to excessive resistance in one or more wires, conductors, etc., either leading up to or exiting from a load (e.g., starter motor); can only be checked with the circuit energized.

Voltmeter A device (usually incorporated into a DMM) used to measure voltage.

Watt A unit of electrical power, calculated by multiplying volts by amps.

Windings (1) The three separate bundles in which wires are grouped in an alternator stator; (2) the coil of wire found in a relay or other similar device; (3) that part of an electrical clutch that provides a magnetic field.

Xenon Headlights High-voltage, high-intensity headlamps that use heavy xenon gas elements.